U0180827

筑味拾年

Ten Years of Flavor Design

郑勇 李涵博 等著

中国建筑工业出版社
CHINA ARCHITECTURE & BUILDING PRESS

图书在版编目（CIP）数据

筑味拾年 = Ten Years of Flavor Design：英汉对照 / 郑勇等著. — 北京：中国建筑工业出版社，2023.3

ISBN 978-7-112-28401-6

Ⅰ.①筑… Ⅱ.①郑… Ⅲ.①建筑设计—作品集—中国—现代 Ⅳ.①TU206

中国国家版本馆CIP数据核字（2023）第032632号

责任编辑：张文胜
版式设计：佘杏奕
责任校对：李辰馨

筑味拾年
Ten Years of Flavor Design
郑勇　李涵博　等著

*
中国建筑工业出版社出版、发行（北京海淀三里河路9号）
各地新华书店、建筑书店经销
北京锋尚制版有限公司制版
北京富诚彩色印刷有限公司印刷
*
开本：889毫米×1194毫米　1/16　印张：12　插页：3　字数：271千字
2023年3月第一版　　2023年3月第一次印刷
定价：**149.00**元
ISBN 978-7-112-28401-6
（40622）

序一

　　"中国建筑之个性乃即我民族之性格，即我艺术及思想特殊之一部"。中华大地幅员辽阔，在五千年的文明发展过程中孕育出了众多灿烂的地域文化，这也造就了中国建筑的多样性。

　　中国建筑西南设计研究院深耕我国西南地区多年，在传承地方文化的建筑行军中已硕果累累。郑勇先生及其团队通过大量走访调研我国西南地区传统的建筑空间，提炼文化要素与现代建筑的美学结合，再转换成建筑语汇指导创作实践，以文化自信、自强的勇气，推动建筑设计的创新和发展，让我由衷地赞赏。其中的一些作品我甚至亲历其中，也略知其设计之匠心，体验过其真实的魅力。

　　室有芝兰气味清，胸有城府天地宽。郑勇先生秉承着"营造百味"的设计理念，深研我国西南地区传统建筑文化脉络，设计完成了众多充满地域建筑文化特色的精品项目，在保留民族传统文化、发掘建筑深层次内涵的同时，还为我们的建筑设计提供了新的思路和角度——中国的建筑设计不仅要面向未来，更要牢记我们背靠五千年的中华文明。

　　随着中国设计市场的开放与发展，这里已成为国际化建筑设计竞争的"战场"。中国建筑师更应该立足于"本土"，聆听"土地"的言语，深入挖掘我们脚下这片沃土中积极、善良、和谐、务实的文化传统，这也将成为中国建筑师在创作过程中绕不开的话题。

　　"城市就是人类文明的产物。人类所有的成就和失败，都微缩进它的物质和社会结构——物质上的体现是建筑，文化上则体现了它的社会生活。"归根结底，郑勇先生所倡导的"营造百味"是建筑物质的呈现，但它同时表达出来的是一种精神。

　　我始终相信，真正有特色的建筑，是因为那里的自然和人文环境有特色，之所以被认可，是因为人们感受到了建筑语汇传递出的文化信息。不必急于去寻找一个大家公认的"中国味"，立足场地、立足场所，创造出一个具体的建筑，改善我们的人居环境，获得更好的生活品质，也许多年以后，大家便会自然而然发现，这就是心中最理想的"中国建筑""川味建筑"。

　　宝剑开锋，十年历练。《筑味拾年》一书是郑勇先生成立工作室十年以来的一次系统总结。无论是晨曦笼罩的民族村落，还是皓月之下的雪山草地，都与他们画笔下的建筑相互呼应。从高楼林立的大街，到烟火不绝的小巷，都是他们灵感的来源，在想象与实践之间，三四月的花，结成了八九月的果。

PREFACE

郑勇先生近年来设计的无论是四川大剧院、盛美利亚酒店，还是成都"东来印象"文化体育中心、大邑长征雪山剧场，都将山光水色融于线条之间、将人文情怀附于框架之上、将民族文化奠于基石之中，通过现代的设计手法，展现出了具有中国特色的建筑风格。每一个设计，从宏观构思到框架腾挪，都倾注了设计师的心血，展现出郑勇先生及其团队对于中华文化的独到理解。

癸卯岁始，大家充满了对于新年的憧憬。如郑勇先生一样，中建西南院人也将继往开来，继续传承和发扬中国优秀的传统建筑文化，创造出更加多元的建筑设计风格。不断在山林流水、乡村部落、城镇街道，点缀下我们对这片土地饱含深情的一笔。

中建集团副总工程师
中建西南院总工程师

PREFACE

序二

当代中国建筑的探索与实践，随着经济的发展、眼界的拓宽及生活方式的变化，呈现出多元的品相和表情。作为国企大院的工作室一员，郑勇工作室承担了走出大院机制缺陷的探索任务及繁荣创作的先行者角色。十年的摸索前行，从团队的里程碑集子中可见努力艰辛的付出，也有丰硕成果的回报。

建筑与其所处的自然与人文环境关系密切，其文化的动态演化属性以纷呈的现实生活方式浸润式地呈现于我们的感知世界中。因此触发了团队创作的"味"作为感知的线索，生发出多样应对环境的设计策略。这本集子记录了工作室十年以来的创作及心路历程，它是大院改革创新的印证，也是具有相同经历建筑师的共情。

十年，一段认知上可长可短的时间。作为一个节点，通过记录工作室团队的创作轨迹，引起对未来的憧憬与思考，激励创作的热情与奋勇，是一种可喜可贺的事情。为此，作为共情者，祝愿工作室的未来，精品不断、人才涌现、大展宏图。

全国工程勘察设计大师
中国建筑首席大师
中建西南院总建筑师

前言

筑味拾年，一晃工作室已经拾年。

孟春三月，天朗气清，坐独椅沉思，抬眼是方寸之地：一两屏荧幕、三四叠图纸、五六杯茶水和七八台电脑，这是十年前工作室刚刚成立时的景象。万事兴起，在由会议室改成的办公室里，工作室开始了喜马拉雅文化中心和中物院成都基地科研楼的设计。当时环境简单，稍显简陋，我虽无"斯是陋室，惟吾德馨"般气定神闲的雅致之情，却也希望在这狭小的办公室能实现建筑师的情怀。

凡事都会经历时间的洗涤，建筑师也一样，我虽祖籍浙江，但因父母早年来到成都工作，我便生在了四川、长在了成都，成了地地道道的成都人。成都人"好吃"，我也一样，从来便喜欢街边小店的纯粹：从老西南院的街道往南出发，在朝阳那一面，有一条人群攒动的小街，全是卖食物的小店，有斜支着四方形的条纹布伞的，有搁着条凳的，有推着小车的，形形色色，杂然并呈。都是小店，不能说大雅端庄，没有餐厅上放块白布然后瓶里插一束花来得体面，但我觉得这里面有生活，有烟火味！

因为喜欢上了成都的烟火味，我在工作室成立之初便带领团队开展地域建筑文化的研究和测绘，开始了我们的寻味之旅。我们探访过成都平原的林盘、桃坪羌寨的碉楼；还记录了川东的吊脚楼、彝族的土掌房……经过了几年深入的接触，发现原来自己对生活了几十年的家乡竟是如此着迷。山光水色与楼宇，身处其中，我为之如痴如醉，何其幸哉？

寻味途中，味归三类：舌尖之味麻辣鲜香，文化之味民韵悠长，人情之味随和包容。我慢慢发现，似乎建筑与味道之间有着某种同样的规则——味道是不同元素之间组合碰撞产生的天作之合，建筑是人们聚散离合间托起的人情冷暖。匠心筑味，设计是不断寻找的过程，在人间滋味中获得了感触，才能寄情于斯、表达于斯。步履不停，寻味不止，如此一来，我们做的建筑也成了有味的建筑。

中物院成都基地科研楼是工作室2012年迎来第一个投标项目。建筑设计首先是功能空间的营造，但我们就希望这幢办公楼不仅仅只有"室"的作用，川人骨子里是悠闲自在的生活方式，喜欢清雅的办公环境，落座之地讲求"心旷神怡、怡然自得"。于是我们设计了九个空中庭院，又用立面肌理表达出四川味道。幸运的是，我们遇到了有情怀的业主，设计方案和他们的想法不谋而合，因此项目中标而且顺利建成，算是工作室成立后的开门红，也算"筑味拾年"的开山之作。

FORWORD

　　缘分总是偶然而来的，2015年底，我遇到了一个非常有趣的小项目：邛泸景区游客中心，它位于西昌著名的邛海风景区的入口处。在项目设计中，我们更多地想如何体现当地彝族的特色。于是，团队一起去了很多彝族的村落。终于在城子村找到了灵感，此间村中，屋隅共山峦一色，自下而上，起伏变化，家家相连、户户相通，一家人的屋顶就是另一家人的晒场，邻犹亲也……这恰恰是村落最打动我的地方，乡亲劳作其间，儿童嬉嬉其间，阡陌贯通其间，自然与人文相互交融。这是彝族人民在悠悠历史中，在大山的怀抱中刻下的自然天成的印记。后来，我们在自己的设计中也保留了这样的印记：我们的建筑呈现出了一种"轻介入"的姿态，整个建筑似从湿地里生长出来与自然融为一体，我们把屋顶高低相连，让市民可以在上面自在游玩、登高望景，重新找到彝族百姓熟悉的生活场景。

　　当味道能够体现文化的时候，就有了味道的沉淀与继承。父亲也是建筑师，我从小耳闻目染受到家庭的影响，从三星堆到锦城艺术宫……设计的点滴都伴随我的成长过程。2019年，四川大剧院建成了，它对我来说难能可贵，并意义非凡。四川大剧院的前身，正是父亲当年设计的锦城艺术宫。在那个百废待兴的年代，我们需要一栋代表性的建筑，鼓舞人们建设家园的热情。秉承这样的赤诚之心，父亲将江碧波老师的"华夏蹈迹"金丝壁画设计在了立面最显眼的位置。所谓踵其事而焕其新生，多年后当我执笔设计大剧院时，也希望能够保留这几幅金丝壁画，老剧院的"气息"当绵延不绝、经久不息。记得父亲当年也经常在茶余饭后讲起设计的点点滴滴，给正在读中学的我留下了很深的印象。在若干年以后，当我开始执笔设计四川大剧院时，才发现这种影响是多么的深刻。

　　以铜为镜，可以正衣冠；以城市为镜，可以知兴替。古老的成都，独特的公园城市，砖瓦相叠间田园风情犹如画卷，有了它新的面貌，老的味道。

　　2015年设计高新体育中心时，有个有趣的细节，令我耳目一新，影响了我平稳已久的设计观。当时和业主一起讨论项目定位，设计团队提出的公园理念得到大家认同，但总觉得差点什么！这时业主一位领导谈了对项目的憧憬：这不是一座高大上的地标，它应该是老百姓喜欢来的体育中心，有欢跃的力量感，也应该充满生活的烟火味！于是，设计随后调整了空间的尺度，开放的公园和宜人的小尺度相结合，公园的活动空间和商业岛穿插在一起。2021年体育中心建成了，世界杯的时候，这里有大屏、有助威声、有交谈声、有孩子和朋友、有满地的啤酒烧烤、有欢快的风吹向远方……这样的场景有浓重的烟火气息，在发展迅速的当下美好得不太真实，这正是公园城市里装载的人情味。体育中心举办了第56届世界乒乓球锦标赛，2023年将承办世界大学生运动会的比赛，我们相信，这里能再现成都公园城市的古老与现代、亲民与时尚。

在中国的文化之中，味是舌之所尝、鼻之所闻，更是情之所藏、文化之所展现。用诗意的"味"来看待建筑，或许建筑更应该是一件艺术品，它不仅是丰富多彩的艺术表达形式，还是记录人情世故、悲欢离合的载体。"味"是精心设计的、是调拌出的酸甜苦辣。前些年在《川味建筑》一书中，我曾试着以"味"来比拟建筑之中的文化内涵、地域特征，这是建筑的性格。而今日所作，为其"味"再添一笔：建筑之中，也有情的风味。食物总是能恰如其分地安慰人心，而建筑之"味"与食物不分伯仲，一箪食、一瓢饮、一豆羹中参悟生活的味之道。不管是建筑本身的风味，还是人寄托在建筑上的人情味，两者相互共生，味又增味，终相辅相成、百味相依。

从事建筑设计三十余年，与千百种美食相遇，与千百个人相识，与千百种人生相知相惜。更有幸在这十年间遇到工作室的小伙伴们，我们因建筑之缘汇聚一堂，聚散之间衍化出独具风味的建筑，又因喜乐相投，关系维系至今长久不坠。这是令人拍案叫绝的命运盛宴，也是令人动容的邂逅。从此岁时存问，相待以礼。

长日碌碌，写作自遣。因"筑"而起，营造"百味"，冠以"筑味拾年"四字，以示所思所悟，矢志不渝。

郑勇
2022年冬

目录
CONTENTS

筑味

寻味

稍长

风味寻源的旅程，永远伴随着偶然和雀跃。

　　"住有所居"是人类最基本的生活要求。每个人的一生，都有关于"家"的记忆。正因如此，人类或因地就势，或顺应自然，倾注了潺潺的情感，发展出灿烂的建筑文化。"寻味"甚似"寻根"。

中物院成都科研创新基地科研综合楼
COMPLEX BUILDING OF CAEP IN CHENGDU

中物院成都科研创新基地，是工作室2012年成立之后中标的第一个项目。虽是一幢常规的办公楼，但我们对设计充满了期待，我们希望通过在建筑中再现某种生活的场景，让这幢建筑变得更加生动。这也成了我们开始"寻味"的起点。

生活场景的再现，与在地的文化密不可分。由于项目地处成都，所以我们试图挖掘出一些富有成都元素的亮点：对于生活在天府之国的成都人来说，物产丰富，从来不会为穿衣吃饭而烦恼。但由于地处盆地，气候湿润、阴雨绵绵，这让人人都对阳光有着特殊的偏爱。但凡有个艳阳天，一定是呼朋唤友，举家出动，找个开敞的"院坝"，喝喝茶、聊聊天。慢慢地，承载人们晒太阳的院落也成为一个象征，变成川西民居最为重要的空间，这里可以汇聚生活、生产几乎所有的活动，而这也正是体现"川味"最好的载体。于是，我们尝试着将"院落空间"作为主题，置入了这座现代化的科研办公建筑之中。

The Complex Building of CAEP in Chengdu is the first competition-winning project undertaken by our studio after its establishment in 2012. Although it is a regular office building, we are full of expectations for the design. We hope to make the building more vivid by reproducing a certain scene of life in the building and it also became the starting point for us to "flavor seeking".

The reproduction of life scenes is inseparable from the local culture. Since the project is located in Chengdu, we also hope to integrate more local elements into the design. For Chengdu people who grow up in the land of abundance, they are rich in products, and never worry about dressing and eating. But as it is located in the basin area, the climate is wet and rainy, which makes everyone has a preference for the sunlight. Whenever there is a sunny day, people go outside with friends and the whole family to find an open "courtyard dam" to drink tea and chat. Gradually, the courtyard for people to bath under the sun has become a symbol and also become the most important space for western Sichuan dwellings, where almost all the activities of life and production can be gathered. It is also the best carrier to reflect the "Sichuan flavor". Therefore, we try to put the "courtyard space" as the theme into this modern scientific research office building.

首先，我们需要完成平地院落与高层建筑的结合，于是我们设计了一个立体化的院落体系：从建筑裙楼开始，以首层的大堂作为起点，人们通过"桥"跨入大堂后，随即可以听到通道两侧的景墙上通过二层裙楼的"水院"缓缓落下的流水声，目光被吸引到二层裙楼的"水院"上时，又可以看到极具地方特色的毛竹景观，水与竹的交相辉映，又通过撒在院子里的阳光，被映衬在了首层大堂斑驳的墙面上。而随着楼层的上移，在总共20层的建筑中，我们以"松、竹、梅、兰"为主题，组织了一条可以"游览"的路径，总共置入了九个各具特色的院落，院落轮廓随着平面功能变化，形成了"曲径通幽"的空间体验。

九个院落分布于整个建筑之上，大约每两层的办公空间都能同时共享一个空中花园，这为使用者提供了一个非常舒适的工作环境。各个院落空间由悬挂楼梯相连，人们可以通过环境宜人的步行系统在整幢建筑中穿行。再加上花园部分采用的都是开放式幕墙，人们可以在休息时间徜徉其中，感受明媚的阳光和徐徐的清风，成都人记忆最深的生活场景在这里得到了再现。

First of all, in order to complete the combination of flat courtyard and high-rise buildings, we designed a three-dimensional courtyard system: starting from building podium, with the first floor of the lobby as a starting point, people go through the "bridge" into the lobby, then they can hear slowly falling water from the channel on both sides of the wall through the second floor podium "water courtyard". When they are attracted to the podium "water courtyard" on the second floor, they can see the bamboo landscape of local characteristics. Water and bamboo add radiance to each other. Through the sunshine sprinkled in the yard, they are set off on the mottled metope in the first floor lobby. And going up with the floor, in a total of 20 layers of building, we take "pine, bamboo, plum, orchid" as the theme to organize a path allowing "tour", with a total of nine distinctive courtyards placed. The courtyard outlines change with the plane function and form the "winding path" space experience.

The nine courtyards are distributed over the whole building. About every two floors of office space can share a hanging garden at the same time, which provides a very comfortable working environment for users. Each courtyard space is connected by hanging stairs, and people can walk through the whole building through a pleasant walking system. In addition, the garden part adopts the open curtain wall, people can even wander in the rest time, feel the bright sunshine and the gentle breeze, and the most memorable life scenes of Chengdu people are reproduced here.

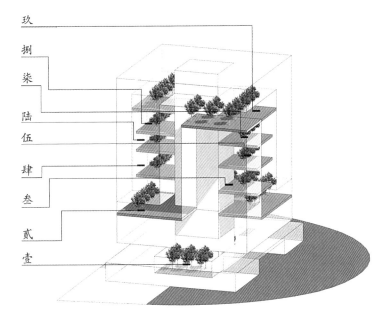

玖
捌
柒
陆
伍
肆
叁
贰
壹

空中云院

云院实景

幕墙实景

　　在感受院落空间的同时，我们在立面设计中也提炼出了传统建筑花窗的式样，让整个建筑的"味道"变得更加丰富。建筑幕墙采用了菱形的单元模块，同时在幕墙玻璃上采用了六种疏密不同的白色彩釉点，营造出不同的韵律感。在空中院落的位置，更是采用开放式的设计，通过调节玻璃间隙的方式平衡通风与美学需求，营造出了中国传统造园中"景致互借"的韵味。

　　While feeling the courtyard space, we also extract the style of traditional architectural flower windows in the facade design, which enriches the "taste" of the whole building. The curtain wall of the building adopts a diamond-shaped unit module, and at the same time, six different dense of white-colored glazed points are used on the curtain wall glass piece, creating a different sense of rhythm. In the place of the sky courtyard, the open design is adopted, and the ventilation and aesthetic needs are balanced by adjusting the glass gap, creating the charm of "borrowing scenery" in the traditional Chinese garden.

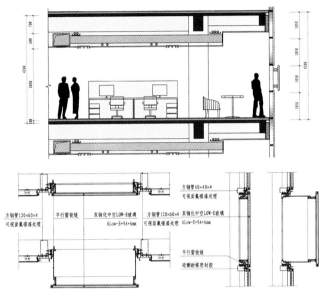

墙身节点

　　在提炼和置入传统元素的同时，我们也力求通过科技手段使"传统"的院落展现绿色生态的现代理念。为了取得更好的通风效果、降低夏季能耗，通过CFD的软件分析在串联的九个空中庭院中增设了尺度适宜的天井，通过与庭院外墙上开放式幕墙的结合，极大提升了建筑内部的"拔风"效果，有效降低了建筑的运行成本。

　　本项目于2013年设计完成，历时3年于2015年竣工。这个项目，设计团队充分发挥了想象力，力求在一幢理性的科研办公建筑中融入温暖的传统文化元素，成为工作室第一个探索建筑与"川味"融合的项目。

While refining and placing traditional elements, we also strive to make the "traditional" courtyard show the modern concept of green ecology through scientific and technological means. In order to achieve better ventilation effect and reduce energy consumption in summer, we added a patio with appropriate scale in the nine aerial courtyards in series through CFD software analysis. Through the combination with the open curtain wall outside the courtyard, the "wind pulling" effect inside the building is greatly improved and the operation cost of the building is effectively reduced.

The design phase was finished in 2013 and took three years to complete in 2015. In this project, the design team fully imagined and tried to integrate warm traditional cultural elements into a rational scientific research office building, making the building become the first project of our studio to explore the integration of architecture and "Sichuan flavor".

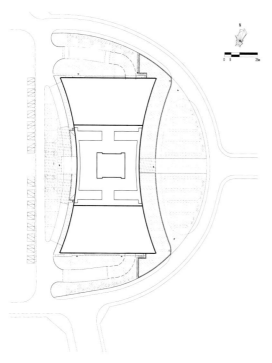

总平面图

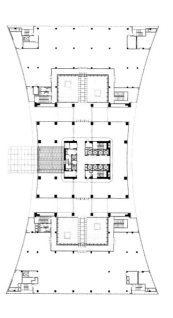

平面图

模型照片 1

模型照片 2

四川大学喜马拉雅文化及宗教研究中心

HIMALAYAN CULTURE AND RELIGION RESEARCH CENTER OF SICHUAN UNIVERSITY

在"寻味"的过程中，四川大学喜马拉雅文化及宗教研究中心是我们在创作实践中的又一次重要尝试。我们在设计中更为深入地挖掘了四川地区民居的传统语汇，通过对庭院空间、屋面造型及细部材料的仔细推敲，期望打造出一个融合地道"四川味"的在地建筑。

本项目坐落在四川大学江安校区内，建筑面积4100平方米，包含博物馆和研究中心两部分功能。作为四川大学内专门研究南亚文化及宗教的机构，建成的研究中心主要研究南亚泛喜马拉雅带国家的文化及宗教，并通过博物馆对其艺术品进行展示。

In the process of "flavour seeking", Himalayan Culture and Religion Research Center of Sichuan University is another important attempt in the practice of our local architectural creation. We have introduced the traditional residential building language of Sichuan into the courtyard space of this building, by virtue of deliberating the courtyard space, roof modeling and fine materials carefully, hoping to create a local building integrating a variety of "Sichuan flavor".

This project is located in the Jiang'an Campus of Sichuan University, with a construction area of 4,100 square meters, including two functions, the museum and the research center. As an institution specializing in studying the cultures and religions of South Asia in Sichuan University, the research center focuses on the culture and religion of the Pan-Himalayan countries in South Asia, and exhibits their artworks through museums.

　　四川民居建筑由于用地限制、外部临街或与其他民居紧紧相贴，大多较为封闭，每个民居环境的打造主要依托自身内部院落，院落成为民居不同功能和布置转换的核心。本项目引入"院落"作为空间主题，设计将博物馆和研究中心分别作为单独体量进行组合，以其形态围合出庭院，各个院落彼此独立却又相互串联，与建筑内部相互渗透，营造出舒适的外部空间环境，形成与功能主体共生的绿色建筑群落。

　　博物馆的两个核心庭院——东院和西园，在具体设计上被赋予不同的空间特点和风景主题，意图以有限面积，造无限空间。东院为了营造出现代博物馆入口空间简洁大气的时代气息，整体塑造更偏向于现代几何的造园手法。相比于东院的仪式感，西园又是另外一番景象——中国园林曲径通幽、步移景异的意境在这里被彰显得更为明显，以自然的"柔"对比呼应了东院现代的"刚"，丰富了空间的节奏与访客的体验，使整个建筑群落成为一个可藏、可游、可观、可赏、可想的景观园林区。

　　Sichuan residential buildings are mostly enclosed due to land use restrictions, external street fronting or close proximity to other residential buildings. The construction of each residential environment mainly relies on its own internal courtyard, which becomes the core of the transformation of different functions and layout of residential buildings. We introduce "courtyard" as the space theme of this project. The concept is to combine the museum and research center respectively as separate volumes and form around the courtyard. Each courtyard is independent of each other but also connects each other. With the internal penetration, the volumes create a comfortable external space environment, forming a green building complex of symbiosis with the functional subject.

　　The two core courtyards of the museum, the East courtyard and the West courtyard, are endowed with different spatial features and scenic themes in the specific design, with the intention to create infinite space with a limited area. In order to create a simple and atmospheric feeling of the entrance space of modern museum, the overall shaping of the East courtyard is more inclined to modern geometric garden making techniques. Compared to the feeling of ceremony of the East courtyard, the West courtyard is with another scene. The artistic conceptions, "the winding path leads to a peaceful secluded place" and "scenes shift with each step taken", of Chinese garden is more obvious here. The contrast of natural "softness" echoes the modern "rigidity" of the East courtyard, enriching the rhythm of space and the experience of visitors, making the whole building community a landscape garden area that can be used to hide, tour, observe, appreciate and think of.

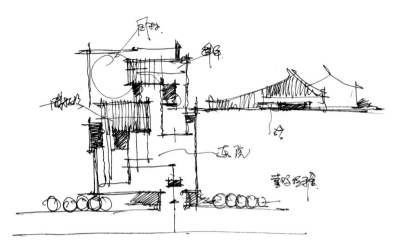

手绘草图

东院

　　民居作为四川最古老的建筑，多采用青瓦坡式屋顶处理，以解决四川多雨季节的屋面排水问题。本项目建筑整体造型采用四川地区传统的坡屋顶符号，通过剪切、变形等手法对其进行现代诠释，使得前后错落、长短不一的坡屋面层层叠叠、高低起伏，形成富有张力的天际轮廓。

　　As the oldest building in Sichuan, the dwellings are mostly roofed with green tiles and sloping roofs to solve the problem of roof drainage in rainy season in Sichuan. The overall shape of the building in this project adopts the traditional sloping roof symbol in Sichuan region, which is interpreted in a modern way by means of shearing and deformation, so that the slope roof with different lengths is layered and high and low, forming a tense skyline contour.

材料细部

材料细部

在四川大学喜马拉雅文化及宗教研究中心材料类型、色彩和质感的选择上，同样延续了四川传统民居用材质朴的特点，原则上尽量屏除浮华的质感与色彩，尽力传达传统民居那种简朴的味道。设计将民居传统的黄色木板墙、灰色青砖、镂空花墙、小青瓦等，通过现代的钢构件、木幕墙、火山岩墙面、金属屋面等现代材料进行重新表达，在色彩及形式构成上与传统文脉形成有效呼应。

佛教作为喜马拉雅带地区人民的重要宗教信仰，几千年来一直以它独有的文化在不断阐释着"人与自然的关系"这一深刻命题。在佛教的语境里，自然与人的关系是如此简单、朴素、和谐，充满着禅学的意味。作为一座地处于四川大学校区内的宗教研究博物馆，建筑设计并未植入大量符号来回应它的主题背景，而是通过现代建筑空间与自然景观的表意系统，挖掘出佛教的精神之光：无论是漂浮在镜水上的栈桥、隐现在草野中的石阶，还是天光倾泻的边庭、黑暗中的转经筒，现代建筑、人工自然、文化展品交相辉映，共同丰润衬托出建筑的精神文化内涵。

In the choice of materials, color and texture of The Himalayan Culture and Religion Research Center of Sichuan University, we also follow the characteristics of simple materials of Sichuan traditional dwellings. In principle, we try to discard the flashy texture and color, in an attempt to convey the simple taste of traditional dwellings. The design re-expresses the traditional yellow wooden wall, gray blue brick, hollow flower wall, small green tile and so on through modern steel components, wood curtain wall, volcanic rock wall, metal roof and other modern materials, forming an effective echo with the traditional context in terms of color and form composition.

As an important religion of people in the Himalayan region, Buddhism has been constantly explaining the profound proposition of "the relationship between man and nature" with its unique culture for thousands of years. In the context of Buddhism, the relationship between nature and human being is so simple, plain, harmonious and full of the meaning of Zen. As a religious research museum in Sichuan university campus, the architectural design has not been implanted with too many symbols to respond to its thematic background, but through the ideographic system of modern architectural space and natural landscape to excavate the spiritual light of Buddhism. Whether it is the trestle floating on the water like a mirror, the stone steps looming in the grass field, or the side court of the sky light pouring down, the tranquil-rotating wheel in the dark, the modern architecture, artificial nature, cultural exhibits add luster to each other, and enrich the spiritual and cultural connotation of Buddhism.

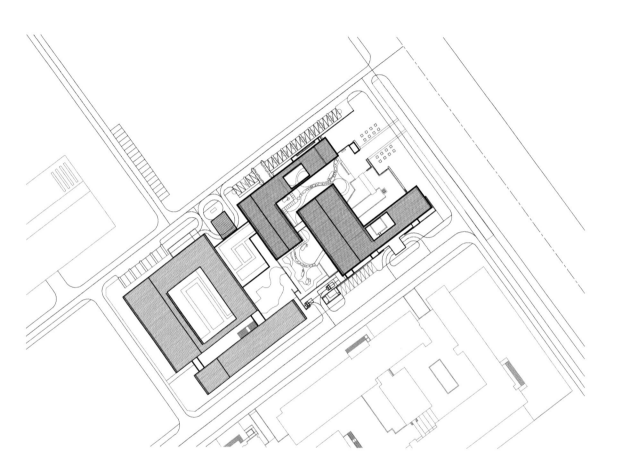

总平面图

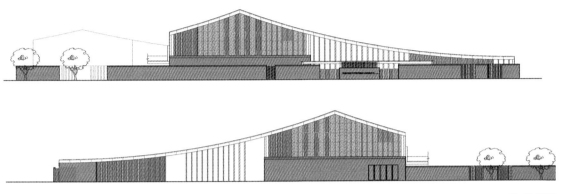

南、北立面图

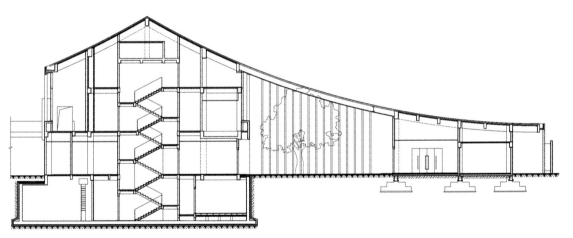

剖面图

博物馆南侧

模型照片 1

模型照片 2

模型照片 3

局部细节

西昌邛海泸山景区游客中心
TOURIST CENTER OF XICHANG QIONGHAILUSHAN SCENIC SPOT

西昌邛海泸山景区游客中心设计始于2014年9月，2018年10月建成投入使用。3000平方米的建筑位于西昌邛海湿地公园入口位置，包含售票厅、餐厅、咖啡厅、土特产售卖、景区入口大门及景区配套服务用房等功能。绵延不绝的湿地、背山面海的环境，使项目场地显得无限广袤和开阔。在这里，大地的张力得以充分体现。面对这样的美景，设计的初衷是创造一个低调的、一个像从大地中生长出来一般、匍匐在地面上的建筑，将真正的空间主角退让给周边的自然环境。

The design of the Tourist Center of Xichang Qionghailushan Scenic Spot began in September 2014, and it was built and put into use in October 2018. The 3,000 square meters building, which includs a ticket office, restaurant, cafe, local specialty sales, entrance gate and supporting service rooms, is located at the entrance of Qionghai Wetland Park in Xichang. The continuous wetland and the environment behind the mountains and the sea make the project site appear infinitely vast and open. Here, the tension of the earth is fully reflected. In the face of such a beautiful landscape, the original intention of the design was to create a low-key building, a building that seems to grow out of the earth, and crawl on the ground, thereby giving back the real protagonist of the space to the surrounding natural environment.

种植屋面肌理

　　有别于在湿地边"建一座房子"，我们更希望塑造一种建筑与湿地共同生长的"景观意向"。充分结合场地高差，建筑轮廓线犹如从大地中生出的岩石一般逐渐显现，有的嵌入水底，有的隐入草地，还有的被芦苇漫过，随着时间，长进土里。整体建筑采用绿色种植屋面，进一步强化了建筑与自然湿地共存的想法。

Different from "building a house" near the wetland, we hope to create a "landscape intention" with the building and wetland grow together. With fully consideration of the attitude difference of the site, the building outline is designed like the rock exposed after the earth cracks, some are embedded in the water bottom, some are hidden into the grass, and some are overflowed by reeds, which grow into the soil over time. The overall building uses a green planted roof, which further reinforces the idea that the building coexists with the natural wetland.

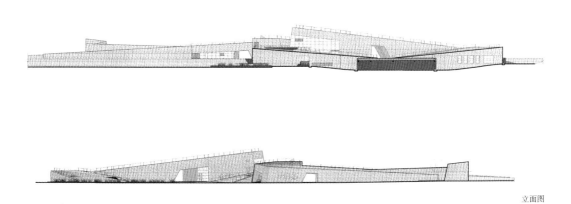

立面图

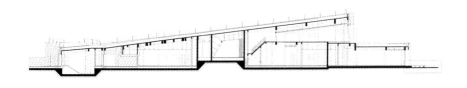

剖面图

城子村钢笔画

　　区别于一种实体的空间构建，"场景感"是这次设计中希望带给使用者的另一重体验。项目位于彝族地区，这里有一种被称为"土掌房"的传统建筑，极具地域特色：成片的民居依山就势而建，或上下相通或左右相连，这家的屋顶就是那家的露台，一户连着一户、一层连着一层。人们利用屋顶晾晒粮食、拉扯家常，生活中的各种故事都在这里发生。而我们也通过现代建筑的空间表达，尽力复原出这种具有活力的生活场景。起伏的屋面结合屋顶公园、观景平台、咖啡休闲商业、景区入园动线等，提供丰富多样的公共交往场所，当使用者行走至建筑坡顶上时，他们更能从高点远眺，饱览湿地和邛海景致。全天开放的公共性，吸引着当地市民及旅游者前来探索和游玩，亦或是进行散步、锻炼、阅读、滑板等活动。在这里，我们看到了各种活动与建筑之间产生的互动，它已经突破了景区游客中心这一单一功能的局限，演变成为一处能够激发公众参与性的城市节点。

　　Different from a physical space construction, the "sense of scene" is another experience that we hope to bring to the users in this design. The project is located in the Yi area, where there is a traditional building called "earthen palm house", which has great regional characteristics: clusters of dwellings are built on the mountain, either connected up and down or connected to the left and right. The roof of the house is the terrace of another house, one household connected with one household, and one floor connected with one floor. People use the roof to dry food, have conversations, and all kinds of stories take place here. We also try our best to restore this dynamic life scene through the space expression of modern architecture. The rolling roof, combined with the roof park, viewing platform, coffee and leisure business, and the scenic spot access line, provides a rich variety of public communication places. When visitors walk on the top of the building, they can look far from the high point and enjoy the beautiful view of wetland and Qionghai Lake. All-day openness to the public attracts local citizens and tourists to explore and play. Visitors can also do walking, exercising, reading, skateboarding and other activities. Here, we see the interaction between various activities and the building, which has grown beyond the single function of the tourist center of the scenic spot to evolve into an urban dynamic node that stimulates public participation.

屋顶测绘

　　彝族"土掌房"多以石材为墙基，用土坯砌墙或用土筑墙，显得厚实朴素。本项目建筑外立面材料力图与这种本土民居形式取得呼应，使得大凉山彝族传统村落中最常见的夯土质感能与建筑的现代形体相结合，以彰显地域特色并且焕发新生。在充分考虑了西昌9度抗震设防、传统夯土的表现力等因素后，我们选择了绿色环保的装饰混凝土板，通过手工设计模板控制混凝土表面的横向肌理，使每块外挂板都呈现出细腻微妙的变化。为了加强建筑形体设计的整体性，装饰板之间的接缝细部被进一步弱化处理，使得整个建筑从远处观望时，表现出岩石破土而出的坚挺、硬朗之势，而当人走近细看和触摸时，又能强烈地感受到夯土温暖、细腻的质感。

　　The "earthen palm house" of the Yi people are mostly based on stone wall, with adobe walls or walls with soil, which is thick and simple. We tried to echo this form of local dwellings with materials of facade of the tourist center building, so that the rammed earth texture, which is the most common in the traditional village of the Yi nationality in Daliang Mountain, can be combined with the modern form of the building to highlight the regional characteristics and rejuvenate. After fully considering Xichang 9-degree seismic fortification, the expressiveness of traditional rammed earth and other factors, we chose eco-friendly decorative concrete light wall panel, controlled the lateral texture of concrete surface by manual design template, so that the texture of each outer wall panel shows delicate and subtle changes in unity. In order to strengthen the integrity of the architectural form design, the joint details between the decorative panels are further weakened, so that the whole building, when viewed from a distance, shows the firm and strong trend of the rock breaking the ground, and when people closely look and touch, they can strongly feel the warm and delicate texture of the rammed earth.

入口水景

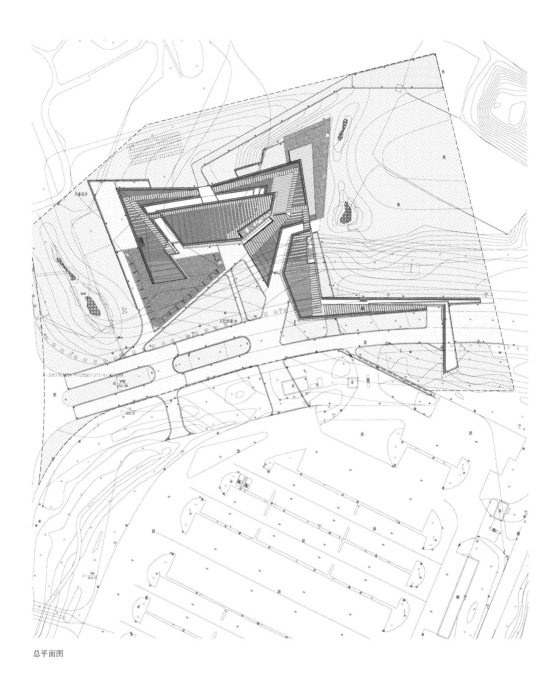

总平面图

屋顶景观

游客中心主入口

四川大剧院
SICHUAN GRAND THEATER

　　成都的天府广场，这片区域自2000多年前建城以来便一直是这个城市的中心。无论是战国的"北少城"、唐代的"唐罗城"，还是明代的"蜀王府"、20世纪50年代的"人民南路广场"，沧海桑田，时光荏苒，这里见证了这座城市数千年来几乎所有重要的历史时刻，成为这座城市最珍贵的历史地标。

　　四川大剧院的设计始于2009年，那时的项目名称还叫作"锦城艺术宫新馆"，用地位于天府广场东侧，原成都锦城艺术宫的旧址。2012年6月，历经3年的设计迎来了最大的修改，用地被调整到了一街之隔的北侧新地块，用地面积也大幅缩水，我们不得不对整个设计进行重新思考。

　　Sinse its establishment 2,000 years ago, the Tianfu Square in Chengdu has been the center of the city. Whether it is the "Beishao City" of the Warring States Period, the "Tangluo City" of the Tang Dynasty, especially the "Shu Mansion" of the Ming Dynasty, and the "Renmin South Road Square" in the 1950's after the founding of the People's Republic of China, it has witnessed almost all the important historical moments of the city for thousands of years in great changes and fleeting time, and has become the most precious historical landmark of the city.

　　The design of the Sichuan Grand Theater took shape in 2009, when the project was still called the "New Jincheng Art Palace", located on the east side of Tianfu Square, the former site of the former Chengdu Jincheng Art Palace. In June 2012, after three years of design, the biggest modification on the design was ushered in, that is, the land use was adjusted to a new land parcel on the north side of the street across, and the land area was also greatly reduced. We had to rethink the whole design.

夜景鸟瞰图

　　虽然天府广场作为成都的城市中心已经延绵了数千年，但由于历史原因，其四周的建筑风格和色彩较为多元和混乱，还没有形成统一的空间秩序。因此我们首先对城市界面进行了梳理：新建筑与已建成的四川省图书馆在城市中轴线上以四川省科技馆为中心左右对称，为了突出科技馆的中心地位，将左右两幢建筑的屋顶高度保持一致，控制在38米。建筑风格也与周边建筑相协调，提取蜀风汉韵的建筑元素，以期唤起人们对于蜀王府的历史记忆。

　　Although Tianfu Square has been the city center of Chengdu for thousands of years, for historical reasons, the architectural style and colors around Tianfu Square are more diversified and messy, without a clear spatial order or distinct spatial characteristics. So we first looked through the urban interface: this newly designed theater and the completed Sichuan Library are symmetrical on the city axis with Sichuan Science and Technology Museum as the center line. In order to highlight the central position of the Science and Technology museum, we kept the roof height of the left and right two buildings consistent at 38 meters. The architectural style is also coordinated with the surrounding buildings, and the architectural elements of Shu style and Han rhyme are extracted in order to arouse people's historical memory of the royal palace of Shu.

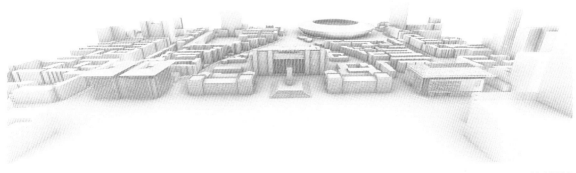

城市空间尺度

新建建筑的规模达到了接近60000平方米，同时容纳一个1600座特大型甲等剧场和一个450座小型甲等剧场，还包括配套的电影院及商业用房。但用地面积仅为11000平方米，非常局促。现状如此，我们也只能在"螺蛳壳里做道场"。经过团队的激烈讨论，最终决定进行一次创新，采用集约化的手法来解决难题。首先在狭小的城市临街面通过架空处理开辟出一个接近1000平方米的室外空间，这里作为"城市会客厅"可以解决剧院的人员集散与动静空间的过渡问题。再通过流线设计，把1600座的大剧场放在了建筑二层，而450座的小剧场则直接重叠在了大剧场的观众厅上方，通过一系列建筑技术的研究与创新，我们成功克服了大小剧场重叠带来的结构和声学难题。

The owner requires the new building area of nearly 60,000 square meters, including a 1,600-seat super-large first-class theater and a 450-seat small first-class theater, as well as a cinema and retail space. However the land area of the new plot is only 11,000 square meters, which is largely insufficient. In view of such a current situation, we can only design the building in a "snail shell". After a fierce discussion by the team, we finally decide to make an innovation and adopt an intensive approach to deal with the problem. First of all, an outdoor space of nearly 1,000 square meters is opened up in the narrow city street surface through overhead treatment. As a "city meeting room", it can solve the transition problem between personnel distribution and dynamic and static space of the theater. Through the circulation design, the 1,600-seat large theater is placed on the second floor of the building, while the 450-seat small theater directly overlaps above the audience hall of the larger theater. Through a series of technical research and innovation, we have successfully overcome the structural and acoustic problems caused by the overlap of the large and small theater.

　　四川大剧院区别于其他城市剧场的最大特点，在于其位于成都这座千年城市的中心，其前身锦城艺术宫还曾是西南地区最高的文艺地标。更为巧合的是，锦城艺术宫的设计者，正是郑勇的父亲郑国英先生，建筑世家的脉络在这个项目得到了传承，新建的四川大剧院也成为锦城艺术宫生命的延续。为了回应天府广场曾经的历史建筑和文化记忆，我们将原锦城艺术宫外墙上由著名美术家江碧波教授以"华夏蹈迹"为主题创作的金丝壁画保留在了观演流线最重要的节点上，让每位经过这里的观众都能感受到新老建筑之间的呼应与对话。

The biggest feature of Sichuan Grand Theater from other urban theaters is that it is located in the center of Chengdu, a thousand-year-old city. Its predecessor Jincheng Art Palace was once the highest artistic landmark in southwest China. More coincidentally, the designer of Jincheng Art Palace, it is Zheng Yong's father, Mr. Zheng Guoying. The context of the architectural family has been inherited in this project, and the newly built Sichuan Grand Theater has also become the continuation of the life of Jincheng Art Palace. In response to the Tianfu Square's past historical buildings and cultural memory, we retain the original gold murals on the exterior wall of the Jincheng Art Palace created by the famous artists professor Jiang Bibo with "Chinese trace" as the theme in the most important viewing node, so that each audience passing here can feel the echo and the dialogue between the old and new building.

鸟瞰图

　　特殊的地理位置，特殊的历史延续，成就了独一无二的四川大剧院项目。从方案设计开始到项目建成长达10年之久，在有限的用地里，我们利用空间关系的叠加，为观众提供了舒适的休闲、观演场所，项目建成后所呈现的艺术效果也达到了预定目标。市民们来到这里，除了享受剧院，还能参与各类艺术活动，亦或喝喝茶、聊聊天，我们在这里再现了多元的文化生活场景，再造了城市文化的新空间。

Special geographical location and unique historical continuation have turned Sichuan Grand Theater into a distinct project. From the beginning of the scheme design to the completion of the project, it lasts for 10 years. In the limited use of the land, we utilize the overlaps of spatial relations to provide the audience with a comfortable leisure and performance viewing place, and the artistic effect presented after the completion of the project has also achieved the predetermined goal. In addition to enjoying the theater, citizens can also participate in various art activities, or drink tea and chat. We can reproduce the diverse cultural life scenes here and create a new space for urban culture.

西广场透视

金丝壁画室内透视

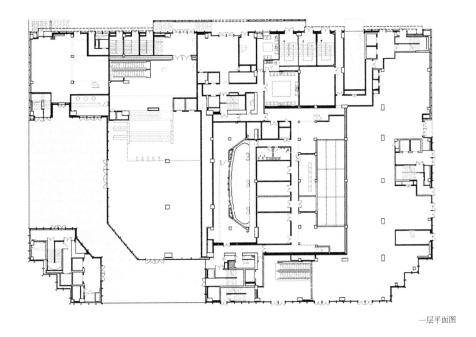

一层平面图

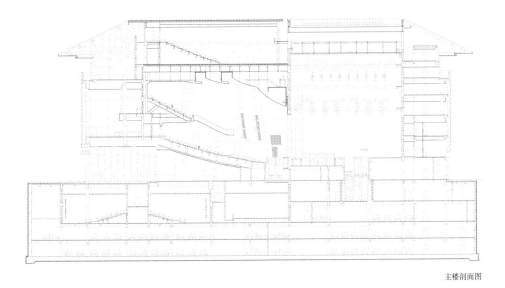

主楼剖面图

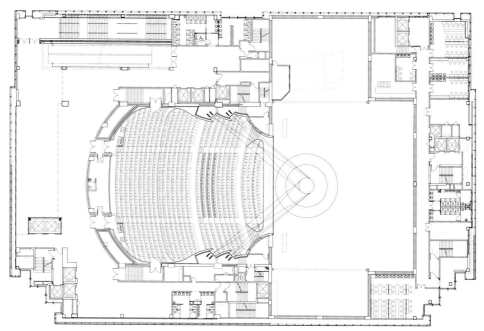

大剧院池座平面图

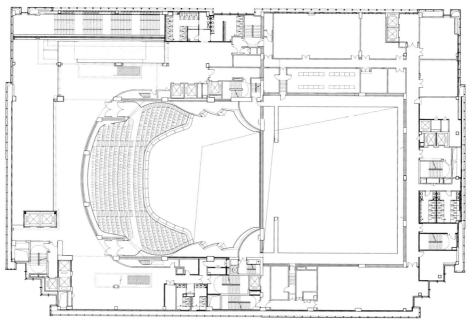

大剧院楼座平面图

大剧场观众厅室内透视

小剧场观众厅室内透视

演出实景

东南角透视

品味

样品

品味是一种品鉴，静心研究，会造就"品味"。

中国建筑文化发展至今，俨然变得多姿多彩，从轮廓造型、空间布局、立面屋顶、设计理念等都有了更新的突破。格物以致知、续往世之绝学，每一代建筑师怀揣赤心、寄情于物，终筑成自己的建筑之"味"。

成都盛美利亚酒店
GRAN MELIĀ CHENGDU

　　盛美利亚酒店，坐落在锦城湖北岸，湖岸草木葱郁，酒店有着城区中绝版的湖景视野。我们在设计中化整为零，采用群体布局的分散策略，让酒店很好地融于锦城湖公园，也隐于公园之中。湖岸远望，丛林掩映，透过树冠间，隐约看到建筑层叠的屋顶，若隐若现。我们希望，建筑以朴实无华的形象示人，却保有厚重的内涵；味淡，却不苍白，是洗净铅华后的真醇；味淡，却不空洞，是繁华落尽后的温润。含蓄、内敛，不咄咄逼人，稍加品味，方能感受到其中的包含的深厚底蕴。

　　The hotel GRAN MELIĀ CHENGDU, located on the north shore of Jincheng lake, with a lush lake shore, it has an one of a kind view of the whole city. We break up the whole into parts, using a decentralized strategy of group layout, so that the hotel is well integrated in Jincheng Lake Park and hidden in it. Far from the lake shore, the jungle, through the canopy, faintly see the layered roof of buildings, faintly visible. With the simple image presented in front of every one, we hope it can keep the deep connotation. It is in a light style but not pale. It has the authentic pureness after great prosperity. It is in a light style but not vacant. It embraces the warmness after richness. Implicate, introverted, and not aggressive, with a little appreciation, you can feel the profound implication contained in it.

项目位于成都，我们当然希望它有着成都的味道。在总体布局上，我们还原了经典的成都院落布置方式，大到官宦府邸，小到百姓民居，这样的院子已经融入进成都人的日常生活中。酒店布局顺应场地朝向，主要功能围绕着中心合院布置；又根据不同功能的特点，设置大小不同的五处院子，筑景造园，移步异景。中心合院及周边建筑采用轴线对称的方式，规整大气，富有仪式感。

酒店采用坡屋顶形式，根据不同高度、不同大小的楼栋形体，形成重檐错落、起伏有致的群体形象。这是我们对成都，也是对古蜀在汉唐历史遗风中的表达，朴实飘逸。建筑采用经典的三段式划分，突出深挑檐、竖线条、实基座等建筑特点；同时，在建筑材料和细节处理上，石材厚重敦实，竖向玻璃窗简洁大气，构件尺度比例精准控制，造型细部体现蜀地建筑的轻盈精巧，营造出酒店厚重大气、典雅精致的现代酒店氛围。我们希望这样"蜀风汉韵"的风格，能让建筑整体形象既有厚实庄重之感，又能延续历史文脉，并传递地域建筑特色。

As it is situated in Chengdu, we naturally want to endow it with a Chengdu style. In terms of the overall layout, we have restored the classical layout mode of Chengdu courtyards, from large official mansions to people's dwellings, and such courtyards have been integrated into the daily life of Chengdu people. The layout of the hotel follows the site orientation, with the main functions arranged around the central courtyard. Meanwhile, according to the characteristics of different functions, we set up five yards of different sizes, constructing landscape and garden, so that people can enjoy different scenes in mobility. The central courtyard and the surrounding buildings adopt the axis symmetrical way, with neat atmosphere and a sense of ritual.

The hotel adopts the form of sloping roof. According to the different height and sizes of the building shape, a double eaves scattered and undulating group image is formed. This is our expression of Chengdu, as well as of the ancient Shu style in the history of the Han and Tang dynasties, simple and elegant. The building adopts the classic three-stage division, highlighting the deep eaves, vertical lines, solid base and other architectural characteristics. Meanwhile, in terms of building materials and details, the stone is thick and solid, the vertical glass window is simple and magnificent, and the component scale proportion is accurately control. The shape details reflect the lightness and fineness of Shu architecture, creating a thick, elegant and delicate modern hotel atmosphere. We hope that such a style of "Shu style and Han charm" can make the overall image of the building have a thick and solemn sense, while continuing the historical context, and conveying the regional architectural characteristics.

内庭院细节

　　作为湖畔酒店，我们在设计上着重强调湖景特色，为客人提供不一样的入住体验。在临湖朝东的大堂落客区，朝向湖面设有叠级的无边景观水池。从落客区回看，远处的湖水与无边水池融为一体，湖水触手可及。湖畔全日餐厅设于大堂楼上，正对湖面；利用宽大屋檐设置的餐厅外摆区，在成都的大部分季节，客人都可以在此品味湖畔闲暇时光。为将湖景资源利用最大化，将客房布置在临湖景观最好的东侧、南侧；同时，利用自然地形高差，对朝南的客房采用吊层处理，增加2层面湖客房楼层，有效提升了湖景客房数量，观湖客房比例约70%以上。

　　锦城湖公园围绕着锦城湖打造，吸引着周边市民，或早起晨练，或饭后漫步，公园已经成为市民生活的一部分。项目用地紧临公园，如何在充分利用景观资源的同时，减小公园和酒店之间的相互干扰，使酒店融入公园，是项目面临的挑战。项目从布局上，沿公园绿道后退30米作为缓冲绿化区，其间按公园绿化标准强化缓冲绿化区的植物搭配，提升公园的绿化环境。另外，建筑化整为零、群体布局的分散策略和建筑形体15米的高度控制策略，也让项目很好地隐于公园之中。

As a lakeside hotel, we emphasize the characteristics of the lakeside scene, to provide guests with a different living experience. In the east, lobby landing area facing the lake, there are stacked endless landscape pools facing the lake. Looking back in the landing area, one can enjoy the distant lake integrating with the boundless pool, and the lake water is within reach. The all-day restaurant is located in the lobby, facing the lake. The extended area utilizing the wide eave of the restaurant allows guests to enjoy the leisure lakeside time for most of the seasons in Chengdu. In order to maximize the utilization of lake view resources, the rooms are arranged in the east and south of the best lake landscape. Meanwhile, the natural terrain height difference is utilized to effectively increase the quantics of rooms which has lake view, thus the number of lake view rooms reaches the proportion about 70% and above.

Jincheng Lake Park is built around Jincheng Lake, attracting the surrounding citizens to get up early for morning exercise, or walk after dinner here. The park has become a part of the citizens' life. The land of the project is close to the park. How to reduce the mutual interference between the park and the hotel and integrate the hotel into the park is a challenge for the project. In terms of layout, the project retreats 30 meters along the park green-way as the buffer greening area, during which the plant collocation in the buffer greening area is strengthened according to the park greening standard to improve the greening environment of the park. Secondly, the decentralized strategy of building seperation into parts, group layout and the height control strategy of 15 meters of building form also make the project well hidden in the park.

内庭透视

　　盛美利亚酒店已于2020年9月开业，其凭借蜀风汉韵的建筑风格、休闲放松的生活场景，迅速吸引了媒体和大众的关注，成为成都新晋的打卡地。项目低调、含蓄地偏居于锦城湖一角，看似平淡，实则是淡而有味，就像薄雾笼罩的远山，看似清浅悠远，里面却藏着万千风景。不禁让人想起一首苏轼的一句词："雪沫乳花浮午盏，蓼茸蒿笋试春盘。人间有味是清欢。"

　　GRAN MELIĀ CHENGDU has opened in September 2020. With the architectural style of Shu style and the relaxed life scene, it has quickly attracted the attention of the media and the public, and has become a new celebrity card in Chengdu. The project is low-key, and implicit. It situates in the corner of Jincheng Lake, which seemingly plain, is light and delicious in fact, just like a mist shrouded in the distant mountains, seemingly shallow and distant, but there are thousands of hidden scenery inside. We can not but think of a Su Shi's line, "Brew a cup of green tea like snow-foam, and taste the spring dish of bamboo shoots in the mountains. The real enjoyment in the world is the taste of light joy".

内庭院透视

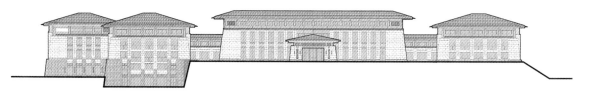

剖面图

内庭透视

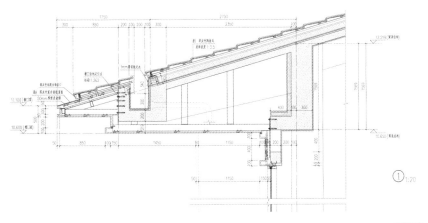

墙身节点

庭院透视

酒店落客区

一层平面图

地下一层平面图

客房

全日餐厅外摆区

成都天府生物产业孵化园一期

CHENGDU TIANFU BIOLOGICAL INDUSTRY INCUBATION PARK PHASE I

　　成都天府生物产业孵化园地处成都市郊，位于依山傍水的自然环境之中。南侧是依托永安湖水库打造的城市公园，东西侧为生物城片区未来的建设用地。颇有雄心的业主和较为开放的规划建设条件让我们能更加自由的创作。项目是一个办公产业园，我们希望利用和打造更加优美的环境，将所有的建筑置于"花园"之中，成为生物产业园践行公园城市理念的一个绝佳示范。

　　Located in the suburb area of Chengdu, Chengdu Tianfu Biological Industry Incubation Park is immersed in the natural environment of mountains and rivers, with the south side standing a city park built near Chengdu Yong'an Lake Reservoir, and the east and west side reserving the future construction land of the biological city area. Ambitious owners and relatively open planning and construction conditions allow us to create more freely. Although the project is an office industrial park, we want to make use of and create a more beautiful environment, putting all the buildings in the "garden", turning it into an excellent demonstration sample for the bio-industrial park to practice the concept of "Park City".

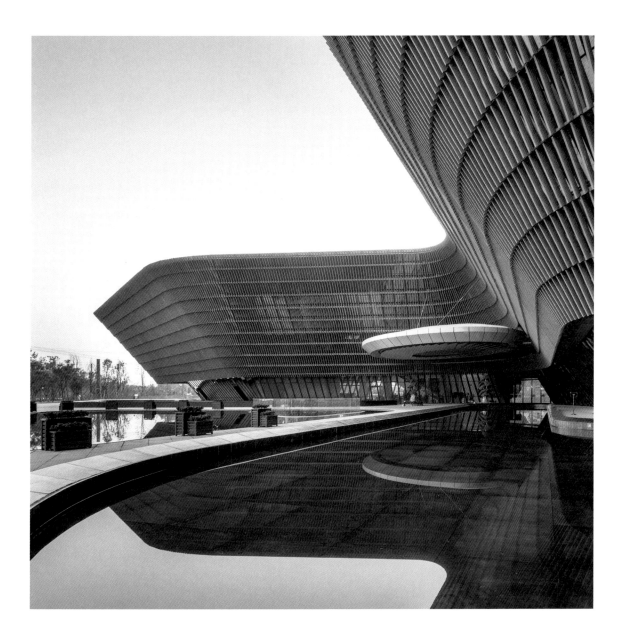

　　我们的方案规划了5个不同形态和空间属性的功能单元，并将它们巧妙地布置在场地内，创造了宽敞的绿色空间和永安湖公园北侧的新景观。其中工程A地块为配套功能区，B、C、D地块为研发办公区，G、E、F地块分别为展示中心、政务、酒店及总部办公区。整体规划包含人才公寓、商业休闲街、研发办公楼、员工食堂等多种功能。错动而高渗透性的空间布局，最大限度地提高建筑的连通性，还可以有效减小热岛效应。而沿中心城市道路南侧区域的景观资源，则被整合成一个城市公园，与隔街相望的永安湖公园串联形成一个嵌入城市肌理的更大的"绿环"空间。

　　项目原规划用地包括8个地块，设计时首先将南侧的3个地块合并，通过整合预留出更大尺度的绿地。而北侧4个地块建筑采取前低后高的布置策略，向园区中心绿带跌落，为各研发楼提供最大化的景观视野。中央公园在用地最南侧，结合场地高差逐渐跌落，通过宽阔的下穿通廊直接与城市干道的另一侧的永安湖公园串联，将公园引入园区，园区链接绿意。

　　建筑体量的错动布置，打通各个地块之间的联系街道，空间得以串联。几个研发地块通过人行的跨街天桥与周边地块联通，从而实现整个孵化园的绿道环线。

　　Our scheme plans 5 functional units with different forms and spatial properties, and puts them cleverly arranged in the site, creating a spacious green space and a new landscape on the north side of Yong'an Lake Park. Plot A of the project is the supporting functional area, and plots B, C and D are the R & D office areas. Plot G, E and F are exhibition center, government affairs area, hotel and headquarters office area. The overall planning includes talent apartment, commercial leisure street, R & D office building, staff canteen and other functions. The staggered and high permeation spatial layout can maximize the connectivity of the building besides effectively reducing the heat island effect. The landscape resources along the south side of the central city road are integrated into a city park, connecting with the Yong'an Lake Park on the south side to form a large "green ring" space embedded in the urban texture.

　　The original planned land of the project includes 8 plots. In the design, the three plots on the south side are first merged to reserve a larger scale green land through integration while the four plots on the north side adopt the layout strategy of low front and high back, falling into the green belt in the center of the park, thus providing the maximum landscape view for each R & D building. The Central Park gradually falls on the south side of the land following the height difference of the site. It directly connects with the Yong'an Lake Park on the other side of the urban road through the wide underpass corridor, introducing the park into the park area, and connecting the park with greenness.

　　The staggered layout of the building volume opens up the connection street between each plot, so that the space can be connected. Several R & D plots are linked with the surrounding plots through the pedestrian overpass, so as to realize the greenway ring of the entire incubator.

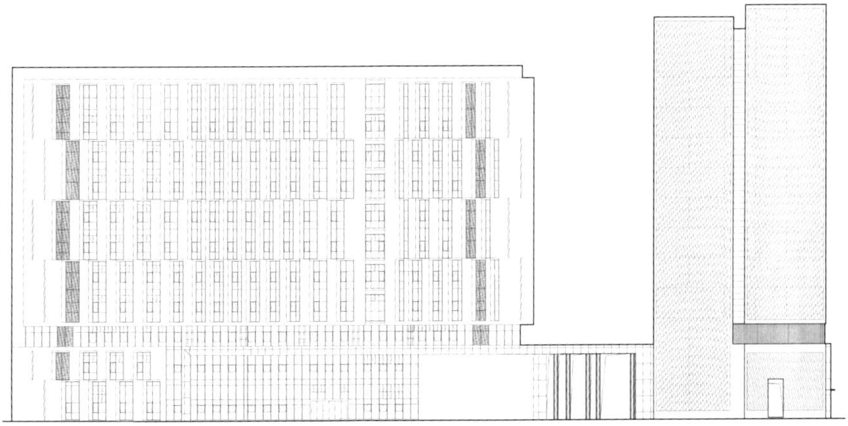

高层研发楼组合立面图

北侧的研发地块建筑在各自片区内围合中心庭院，局部设置丰富的下沉庭院，成为各个地块的空间核心；南侧的总部办公塔楼旋转角度布置，在获取最大的景观视野的同时，让建筑灵动地嵌入中央公园一侧，仿若森林里的园区。

不同的地块在各个院落、建筑群落空间中，控制空间界面，打造近人的空间体验感，削弱产业园建筑的秩序感和体量压力。无论是架空门廊、镂空砖墙，亦或是错动格栅、出挑平台，皆是为打造园区内部的宜人尺度感。

小墙砖、哈佛红，这种外墙肌理与色彩是学院建筑最典型的符号。高低错落的研发楼、红白相间的立面色调、结合清水混凝土涂料的门头和金属构件的立面元素点缀，更辅以底层镂空砖的肌理变化，打造出原汁原味的学园感受。无论走在街巷中，矗立在院落里，抑或是走过树荫下，漫步在天桥上，映入眼帘的建筑场景都能勾起着每人心中校园时代的回忆。

The development plot buildings on the north side have enclosed the central courtyard in their respective area, with rich sunken courtyards to become the spatial core of each plot. The south side of the headquarters office tower is arranged at a rotating angle to obtain the maximum landscape view while making the building flexibly embedded in the side of the central park, just like a park in the forest.

Different plots are in the landing space of each courtyard and building complex, controlling the space interface to create a sense of spatial experience of loved ones, and weaken the sense of order and volume pressure of the industrial park building. Whether it is an overhead porch, a hollow brick wall, a staggered grille or a cantilever platform, they are all designed to create a pleasant sense of scale inside the park.

With small wall brick in Harvard red, the exterior wall texture and color is the most typical symbol of college architecture. The high and low R & D building, the red and white facade color, combined with architectural concrete paint door head and metal components of the facade elements embellishment, supplemented by the texture changes of the underlying hollow brick, generate the original academic garden experience. Whether walking in the streets, standing in the courtyard, wandering through the shade of the trees, or strolling on the overpass, the architectural scene that bumps into sight can recall the memories of the campus era in everyone's heart.

在强调总体公园化的同时，研发办公楼的平面设计依然是理性和科学的。设计采用辅助空间分置两侧的布置方式，既可单层整体使用，也可拆分成两间独立的办公空间。单元办公面积以150平方米为模数，可根据企业需求灵活拆合，逐级扩大。生物研发产业所需要的通风及空调设备平台被设计在建筑端部山墙，结合左右反斜的竖矩管百叶进行有效美化，在满足功能需求的同时解决美观问题，成为研发楼单体的一大特色。实验室所需要的通风管道，则组织在竖向室外井中，整体连接至屋面排放。

While emphasize the overall landscaping of the site, the plan design of R & D office building is still rational and efficient, it adopts the plane mode of separate placement on both sides of the service space, which can be used in one floor as a whole, and can also be divided into two independent office spaces. The office area of a single unit is 150 square meters, which can be flexibly dismantled, closed and scaled up step by step. The ventilation and air conditioning equipment platform needed by the biological R & D industry is designed on the gable at the end of the building. Combined with the left and right reverse inclined vertical rectangular pipe louvers to effectively beautify while meeting the functional needs and solving the aesthetic problem, which has become a major feature of the single R & D building. The ventilation duct required by the laboratory is organized in a vertical outdoor well and is collectively connected to the roof for emission.

绿林中的文化中心

　　以项目的中央公园为核心打造的园区，不仅仅服务内部的研发、办公人员，更是城市空间的重要节点。永安湖北岸这片红砖绿影的复合园区建成以后，随着成都生物产业的快速兴起，也逐渐演变成了充满活力的重要城市功能区。

This project with a core central park is not only served for internal research and development and office personnel, but also an important node of urban space. After the completion of this complex park on the north bank of Yong'an Lake with the red bricks and green shadows as well as the rapid rise of the biological industry in Chengdu, it will eventually evolved into an important urban functional area full of vitality.

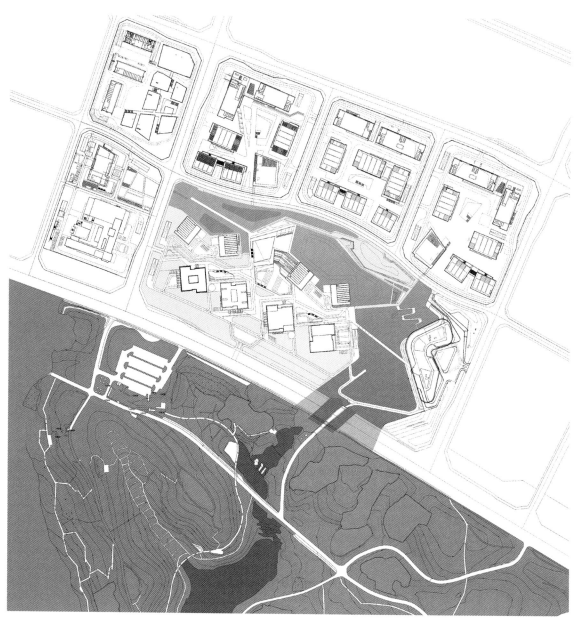

总平面图

孵化园透视

门头透视

永安湖景观透视

文化中心局部

下穿隧道透视

A 地块组合立面

成都高新体育中心
CHENGDU HIGH-TECH ZONE SPORTS CENTER

　　成都高新体育中心位于成都高新区，项目总建筑面积201600平方米，包含多功能体育馆、全民健身馆、综合服务中心、体育场、垒球场、风雨球场等多种功能，其中的多功能体育馆具有座席12900余座，达到特大型甲级体育馆标准，可以满足篮球、排球、体操、冰球等国内国际单项比赛的要求，同时也能满足文艺表演、会议会展等多功能的需求。

　　Located in Chengdu High-tech Zone, the Chengdu High-tech Sports Center project consists of a total floor area of 201,600 square meters, including multi-functional stadium, public gymnasium, comprehensive service center, softball stadium, outdoor sports courts and other functions. Among them, the multi-functional stadium is equipped with more than 12,900 seats, reaching the standard of a super-scale A-class gymnasium. It can both meet the demands of domestic and international competition such as basketball, volleyball, gymnastics, ice hockey and other sports, and the needs of artistic performance, conference and exhibition.

成都高新体育中心自建成以来，已承办第56届国际乒联世界乒乓球团体锦标赛，并将于2023年承办第31届世界大学生夏季运动会，是成都提出"世界赛事名城"建设目标后投入使用的第一批运动场馆。

项目设计始于2014年，彼时的项目所在地成都高新中和片区还是以居住社区为主，用地周边环境品质并不理想，缺乏相应的公园和体育活动设施，无法满足现代市民生活的需求。基于此背景，项目团队希望塑造一个新的城市区域地标，提升在地的环境品质及人文气质，将城市建筑、运动健康与绿色生态环境形成更好地融合，带动周边地区整体生活品质的提升。

Since its completion, Chengdu High-tech Zone Sports Center has held the 31st Summer Universiade, and the 56th ITTF World Team Table Tennis Championships. It is among the first batch of sports venues put into use after Chengdu government put forward the construction goal of "World Competition City".

The design of the project began in 2014 when the project location in Zhonghe Area, Chengdu High-tech Zone was still mainly for residential communities with poor surrounding environment quality around the plot, and insufficient corresponding parks and sports facilities, failing to meet the needs of life for modern residents. Based on this background, the project team hopes to create a new urban regional landmark, improve the local environmental quality and humanity temperament, in order to better integrate urban infrastructure, sports health and green ecological environment, and stimulate the improvement of the overall quality of life of the surrounding area.

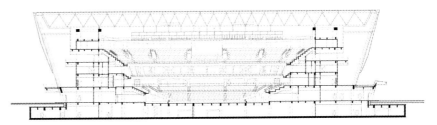

综合体育馆剖面图

　　我们以"创造活力公园"为规划设计的初衷，充分剖析建筑与人、环境、城市的关系，通过在园区内打造生态绿色公园，吸引人流，聚揽人气，并将"公园"作为城市功能的补充。整体规划将主体建筑场馆相对集中布置，让更多土地资源用于景观绿化和室外场地的打造，创造绿树成荫、通透舒缓的生态体育公园。围绕中央体育场的环形景观轴，引入城市人流，延伸至西侧用地外城市绿地，形成延续的城市生态绿轴。用地东侧布置风雨球场及室外场地，以抬升的建筑体量结合侧面草坡的生态消隐，阻挡场地附近高速路的噪声、烟尘对场地产生的负面干扰，并通过建筑抬升形成场地高差。整个用地强化高低起伏的草坡公园景观，结合6米标高的裙房屋面，通过平台、踏步、坡道、连桥形成不同标高步行路径，让室内外各个功能空间自然衔接，人们从户外可到达建筑和室外活动场地的各个层面，丰富健身体验。

　　With the original intention "creative vitality park" of the planning and design, we fully analyze the relationship between architecture and people, environment and city, and create an ecological green park in the center to attract people, gather pedestrain circulation, and take the "park" as a supplement to urban functions. The overall planning of the main building venues are in a relatively concentrated layout, so that more land resources can be used for landscape greening and the outdoor venues, and to create a tree-lined, transparent and comfortable ecological sports park. Around the circular landscape axis of the central stadium, the urban flow is introduced and extended to the urban green space outside the land on the west side, forming a continuous urban ecological green axis. The outdoor sports courts are arranged on the east side of the plot, with the volume of the raised building and the ecological elimination of the grass slope on the side to block the negative interference from the noise and dust of the highway near the site. Besides, the height difference of the site is formed by lifting the building. The whole land emphasizes the undulating grass slope park landscape, combined with the 6-meters elevation podium roof. Different elevation walking paths are formed through platforms, steps, ramps and bridges, so that each indoor and outdoor functional space can be naturally connected, and people can reach all levels of buildings and outdoor activity venues to enrich the fitness experience.

　　设计将体育运动与休闲娱乐及衍生商业配套功能统筹设计，提出了公园里的"商业岛"概念，希望能在这里体现出成都的"人情味"。大尺度的公园、体育设施与小尺度的商业、休闲相结合；公园的活动空间与商业的活动空间相结合，打造出浓浓的"成都味"，实现体育中心的"自我造血"，创造出经济效益。

The design sets sports, leisure, entertainment and commercial supporting functions as a whole, and a concept of "Commercial Island" was conceived in such a sports park area, in order to represent the "human warmth" here of Chengdu citizen. The large scale park, sports facilities and small scale retail space combine together, and the leisure space of both park and commercial space combine together to create a strong "Chengdu Flavour", as well as to realize the "self-metablolism" of the sports center and create economic benefits.

沿街透视

　　四川的"蜀锦"源远流长，自身勾勒出的光泽肌理集传统艺术和现代审美于一体。高新体育中心的外立面设计强调建筑群体的整体性，通过引入流畅的曲线和不同穿孔率的金属穿孔板，形成疏密有致、舒缓灵动的建筑表皮，对四川传统的"蜀锦"肌理进行抽象表达，演绎出独特的现代建筑形象，使之既有独特的艺术风格，又能传递浓浓的四川文化情愫，体现成都作为"锦官城"所特有的雅致、时尚而又个性的生活格调。"蜀锦"的外立面肌理进一步融合"韵动丝路、绿绣锦城"的理念，设计希望建筑如蜀锦缎带一般跃动于锦官城中，带动起一片充满活力的城市绿洲。

The distant-origin "Shu brocade" from Sichuan shines the luster texture outlined by itself integrating traditional art and modern aesthetics. The facade design of the Chengdu High-tech Zone Sports Center emphasizes the integrity of the building group. Through introducing smooth curve and different perforating rate of metal perforation plate, the even, soothing, and vivid building surface is formed, expressing the Sichuan traditional "Shu brocade" texture in an abstract manner, depicting a unique modern architectural image, giving it both unique artistic style and profound Sichuan culture, and reflecting the unique elegant, fashionable and individual life style of Chengdu as a "Jinguan City". The facade texture of "Shu brocade" further fuses the concept of "Rhyme lights Silk Road and green embroiders Jincheng". By virtue of the design, it is hoped that the building will radiate like Shu brocade ribbon in the city, and drive a vibrant urban oasis.

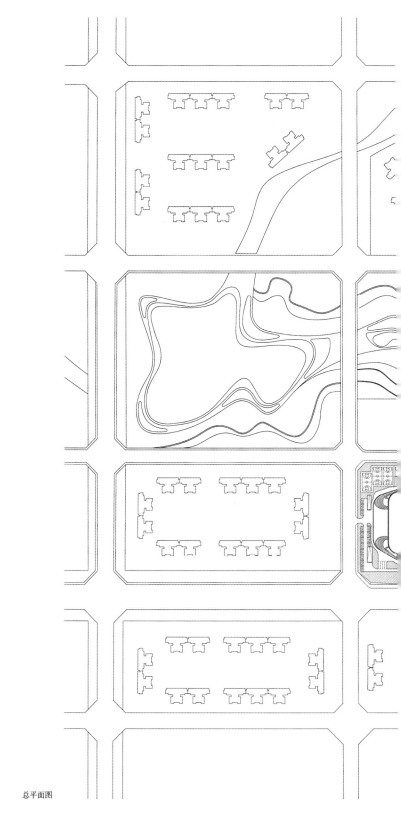

总平面图

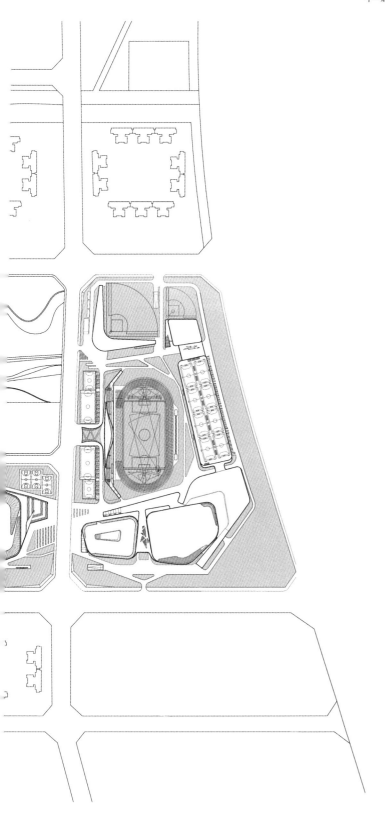

公园侧俯瞰

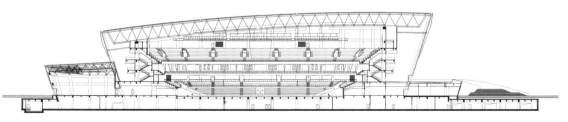

体育场剖面图

东南侧鸟瞰

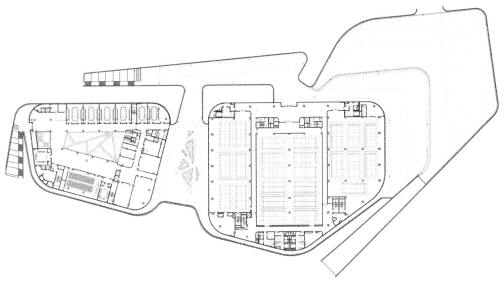

全民健身中心及配套服务用房二层平面图

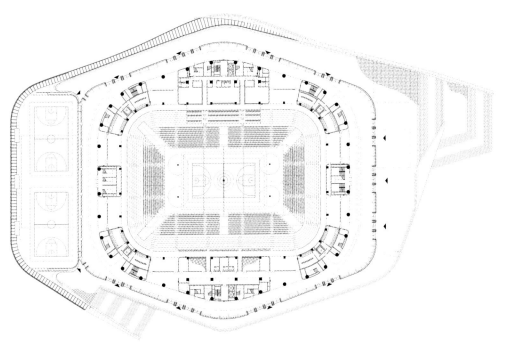

多功能体育馆二层平面图

休息厅透视

内庭透视

多功能体育馆场内透视

四川名人馆
SICHUAN HALL OF FAME

　　如何更好地将建筑的"型"与"意"进行结合，这是我们"寻味"多年之后开始思考的一个新课题。2021年，工作室参与的四川名人馆项目给了我们一个设计实践的机会。

　　和我们近年来的很多项目一样，四川名人馆的用地位于四川天府新区的CBD核心区。但这个项目的特别之处在于其项目用地紧临CBD内的成都天府公园，这是一个拥有3450亩用地的超大城市中央公园。对于我们来说，需要更加谨慎地处理建筑和环境的关系，不致于让这个规模约52000平方米的建筑与环境失衡。同时，这个项目又是一个以展示四川历史名人、文化脉络为主要功能的博览建筑，我们也希望它能在所处的这个大环境中，更加恰当地表现出四川本地的文化内涵。

　　We began to think deeply on a new topic after "seeking flavor" for years, that is, how to better combine the "shape" and "meaning" of buildings. It was the project of Sichuan Hall of Fame museum that our studio participated in 2021 that allowed us to have the opportunity to practice.

　　Similar to many of our projects in recent years, Sichuan Hall of Fame museum is located in the CBD of the Tianfu New Area in Chengdu, Sichuan. One of the special parts of this project is that the plot is closed to the Chengdu Tianfu Park in the CBD, a super large city central park covering an area of 3,450 acres, making it more necessary for us to cautiously deal with the relation between the building and the environment, so as not to imbalance the building of about 52,000 square meters and the environment. Meanwhile, in view that it is an expo architecture functioning in displaying Sichuan historical celebrities and cultural context, we hoped to more appropriately present the local cultural connotation of Sichuan in it and in this environment.

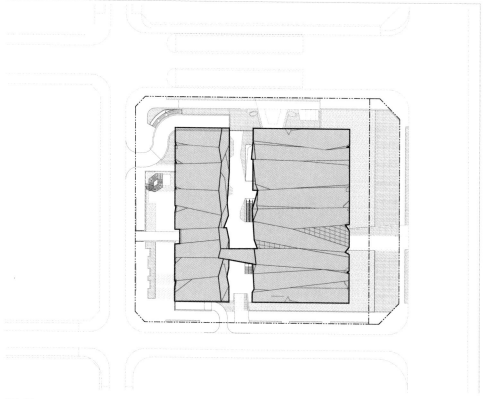

总平面图

　　蜀山——四川因山得名，绵延逶迤。蜀水——蜀地内江河纵横，钟秀灵毓。蜀山、蜀水营造出四川独特的地理自然环境，铸造和形成了四川人特有的精神风貌和个性特征。于是，我们决定从四川的山水中汲取灵感，首先以"蜀山绿谷"作为设计理念，引入"一山两谷"的形态和空间格局，再通过切削建筑体量的方式，塑造抽象的蜀山形态。

　　Sichuan is famous for the stretching and continuing mountains (the short name of Sichuan is Shu). Within the region of Shu, there are rivers in length and breadth called Shu Rivers, which condense the aura between heaven and earth, and nourish excellent characters. It is exactly the Shu Mountains and Shu Rivers that have created a unique geographical and natural environment of Sichuan province, cast and formed a unique spirit and personality characteristics of Sichuan people. In this regard, we are inspired by the landscape of Sichuan. First of all, taking "Shu Mountains and Green Valleys" as the design concept, we introduced the form and space pattern of "One Mountain and Two Valleys". Then, by virtue of cutting the building volume, we shaped the abstract form of Mount Shu.

南侧透视

　　在建筑立面肌理的选择上，我们采用了四川地区最常见的红色砂岩石。这是四川盆地最常见的一种岩石，较易风化，风化后发育形成的紫色土又是一种富含矿物养分的自然土壤。在某种意义上来说，正是红砂岩造就了"天府之国"的千里沃土。同时，由于红砂岩易于开采，历史上人们也常常将其用作建筑材料。正是因为这些原因，我们以红砂岩作为骨料试制了一种新的混凝土外墙材料，不仅利用其"红色"的外表唤起了一种具有四川味道的建筑记忆，同时，也利用材料本身的特性形成了"山石"的具象表现，诠释出蜀山丰富的细节与灵动。

　　In terms of selecting materials for the architectural facade texture, we adopted the most common red sandstone in Sichuan region. In the Sichuan Basin, it is the most familiar rock. It is easy to be weathered. After weathering, it will be developed into dark purple soil, which is a natural soil rich in mineral nutrients. In a sense, it is the red sandstone that has created the fertile soil of thousands of miles in the "Land of Abundance". Furthermore, for the convenience of red sandstone exploitation, people often used it as a building material throughout the history. On account of these reasons, we create a new facade material based on the red sandstone, thus not only making use of its "red" appearance to arouse the architectural memory with the Sichuan style but also utilizing the characteristics of the material itself to form the concrete presentation of "mountain stones" and interpreting the ample details and ethereality of the Mount Shu.

在空间环境的表达上，设计延续公园城市的理念，通过首层建筑架空将天府公园的绿色引入场地之内、建筑之中。场地内规划的生态绿谷，绿林成荫。既融合了项目中名人展示馆与文化体验中心两个不同的功能业态，也为二者提供了立体展示和休闲的场所。绿谷内侧界面好似被剖开的玉石内部，晶体罗列，形态万千，为展厅参观和城市公共空间带来移步异景的流线体验。

我们还在展馆内部打造"名人星谷"，通过设置开敞明亮的中庭，将建筑内部的绿谷和天府公园进行空间和视线上的连通，进一步强化公园城市和建筑内部的生态对话。川人对蜀道的探索与开拓由古至今，从未停歇。结合"星谷"中庭空间，设计引入"蜀道"的概念组织参观动线，整个行进体验也融入了古蜀人进出蜀地的独特体验。"星谷"朝向东侧天府公园，拥有通透的玻璃景窗，使人在观展之余可以感受公园城市的独特魅力。

Referring to the expression of the space environment, the design continued the idea of the park city. The ground floor of the building was lifted up to introduce the green land of Tianfu Park into the site and the building. In the site, there was an existing ecological Green Valley with green forest shade, which both integrated the two different functional formats of celebrity exhibition hall and cultural experience center, and provided a three-dimensional display and leisure place for them. Moreover, the inner interface of the Green Valley was like the interior of a jade being exposed by splitting, with various crystals listed and myriad forms, bringing a streamlined experience of different scenery for the exhibition hall and the urban public space.

We also created a "Celebrity Star Valley" in the museum. The bright and wide atrium in the building spatially and visually connected the Green Valley and the Tianfu Park, which further strengthened the ecological dialogue between the park city and the building interior. Since the ancient times, the Sichuan people have never stopped exploring and developing Shudao (roads leading to the Shu region). As a result, combining the atrium space of the "Star Valley", the design introduced the concept of "Shudao" to organize the dynamic visit line. In the entire visiting process, the unique experience of the ancient Shu people entering the Shu region was also interpreted. Besides, the side of the "Star Valley" facing the east Tianfu Park was installed with transparent glass windows for viewing so that people could enjoy the distinctive charm of the park city in addition to going around the exhibition hall.

绿谷透视

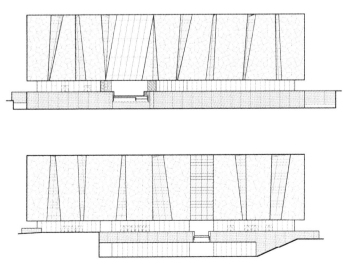

绿谷立面图

西北侧透视

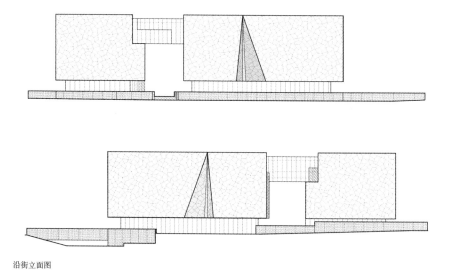

沿街立面图

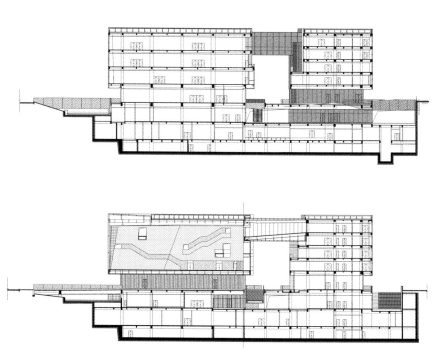

沿街剖面图

造味

和散

建筑的背后，是匠人之心。

　　人们南来北往，行色匆匆，若干不甘平凡的人们依旧在忙碌，依旧在创造"新味"。而如此千滋百味的建筑世界，每一个设计师都担负着不同的调味重任。承上启下、更古增华、科学创新、同造百味。

成都"东来印象"文化体育中心
CULTURE AND SPORTS CENTER OF ORIENTAL IMPRESSION, CHENGDU

成都"东来印象"文化体育中心项目设计开始于2017年7月，2022年3月建成。项目功能几乎涵盖了城市文化体育配套的各种类型，是一个综合性很高的文体中心项目，包括"八馆三中心两园一家一大剧院"：图书馆、文化馆、美术馆、博物馆、档案馆、地方志馆、体育馆、游泳馆；广电中心、青少年活动中心、全民健身和国民体质监测中心；体育公园、文创产业园、文艺之家、大剧院等，规划建筑面积达到419900平方米。设计采用多元功能立体复合的方式，打造一个充满"蜀味"的城市绿色公园，将市民的文体活动融入城市生活之中。

The design of Culture and Sports Center of Oriental Impression, Chengdu project started in July 2017. The project was built in March 2022. The function of the project almost covers the various functions of urban culture and sports. It is a highly comprehensive cultural and sports center project which includes "eight venues, three centers, two parks and a large theater": Library, Cultural center, Art gallery, Museum, Archives, Local Chronicles hall, Gymnasium and Natatorium; Radio and Television Center, Youth Activity Center, National Fitness and National Physical Fitness Monitoring Center; Sports Park, Cultural and creative Industrial Park; Literature and Art Home; Grand Theater, etc. Moreover, the planning construction area reaches as many as 419,900m^2. The design adopts a multi-function and three-dimensional comprehensive way to create a city green park full of "Shu flavor". It integrates the citizens' cultural and sports activities into the city life.

在自然生态策略上，我们响应成都"公园城市"号召，希望将这个项目打造成为一个充满"蜀味"，并且与天府绿道接驳的"文体公园"，与沱江、鳌山共同融入天府绿道的城市格局，同时让建筑与公园融为一体。为实现"外环内心"的双层公园结构，规划上建筑沿周边布局，留出150亩内心公园，公园景观打造颇具蜀味，纵横交错的平面肌理与运动场地相融合，显现出川西林盘的风貌神韵；开和有序的景观水体，配合以跌水、镜水等多层次的表现方式，与建筑相辅相成，呼应川人乐水的精神境界。建筑退让四周城市道路30~50米，构成外环公园，打造银杏长廊与绿道空间，各单体公众入口朝向内心，后勤入口朝向外环，实现人车分离。

"活力环"的融入加强了多功能复合的整体营造。结合绿道系统，架空的"活力环"串联南北地块的各栋建筑：上层为"活力带"，与各建筑的共享边庭、入口大厅有效互动，实现室外与室内的有机交融；下层为休闲带，引入"服务岛"，打造半室外商业活动空间，加强场馆、公园与配套商业的互动，吸聚周边人气，以馆养馆。"活力环"将各个文体博览功能有机串联在一起，同时将南北两区域紧密联系起来，形成融入城市生活的一体化多功能城市文体集群。

In terms of natural ecological strategy, to response to the call of "Park City" of Chengdu, we will create a "Cultural and sports park" which is full of "Shu Flavor" and connected with Tianfu Greenway. The project integrates into the urban pattern of the Greenway of a self-sufficient city together with Tuojiang River and Aoshan Mount, and the buildings integrated with the park. In order to realize the "outer ring inner core" two-layer park structure, buildings are placed along the surrounding layout, leaving a 150 acres inner park. The landscape dresses the park with Shu flavor, with criss-crossing plane texture and sports field integration, showing the western Sichuan Linpan style charm. The open and orderly landscape water bodies, with the multi-level expression of falling water and mirror water, complement each other with the architecture, echoing the spiritual realm of Sichuan people's love of water. Around 30 to 50 meters of setback space is made between buildings and surrounding urban roads, creating an outer ring park, and a Ginkgo promenade and greenway space. Each single public entrance connecting the interior, and the logistics entrance connecting the outer ring can achieve the separation of people and vehicles.

The integration of "vitality ring" strengthens the overall creation of multi-functional composite. Co-working with the greenway system, the overhead "vitality ring" connects the buildings of the north and south plots. The upper level is the "vitality zone", which interacts effectively with the shared side court and entrance hall of each building to achieve the organic blend of outdoor and indoor. The lower level is a leisure zone. The "service island" is introduced to create a semi-outdoor commercial activity space which strengthens the interaction between venues, parks and supporting commercial space, attracts the surrounding pedestrains, and supports the museums. The "vitality ring" organically connects the functions of various sports and exhibitions together. And, at the same time, closely connects the north and south regions, form an integrated multi-functional urban sports and sports cluster integrated into urban life.

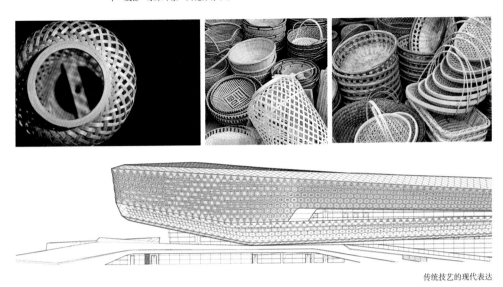

传统技艺的现代表达

为了让建筑与景观相映成趣，我们希望将建筑立面处理得更加通透。成都地区传统的瓷胎竹编技艺为我们提供了绝佳的设计灵感，这也成为建筑形象设计中很好的文化表达。设计将"瓷胎竹编"的意象进行抽象简化，生成参数化表皮，表皮顺应建筑造型进行变化，好像于建筑表面流动的浓淡笔墨。此时的建筑更像是瓷胎主体，"竹编表皮"浮于其外，通透灵动。建筑依靠"竹编"这一半透明介质，让室内空间与周围的城市公园不再是直白的对立关系，两者在交流互动中演变出无限的可能。

In order to perfectly blend the building and landscape together, we make the facade more transparent. The traditional bamboo weaving technique of porcelain in Chengdu provides us with excellent design inspirations, which has also become a great cultural expression in the building appearance design. The design abstracts and simplifies the image of "porcelain and bamboo weaving" to generate a parametric skin that changes according to the building shape, like the thick and light ink flowing on the surface of the building. At this time, the building is more like the main body of porcelain, and the "bamboo woven skin" floating around its outside, transparent and flexible. The building relies on the transparent medium of "bamboo weaving" to make the interior space and the surrounding urban park no longer a direct opposite relationship, and thus the two have evolved infinite possibilities through communication and interaction.

　　成都"东来印象"文化体育中心的总体布局和建筑设计，始终贯彻"公园城市里的公园，公园中的文体中心"这一理念，将建筑与景观生态紧密结合，使得绵绵不断的城市活力不断向外延伸，创造出东来成都最具地方特色的"蜀味"表达。

　　The general layout and architectural design of Culture and Sports Center of Oriental Impression, Chengdu carry out the concept "park in the park city", "cultural and sports center in the park". The close combination of architecture and landscape ecology, with its continuous urban vitality extending outward, creats a unique expression of "Shu Style" and regional characteristics here in Chengdu.

北区活力环透视

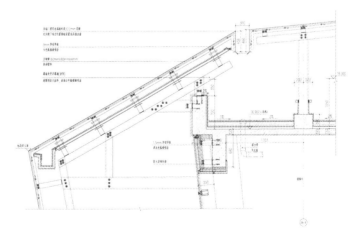

墙身节点

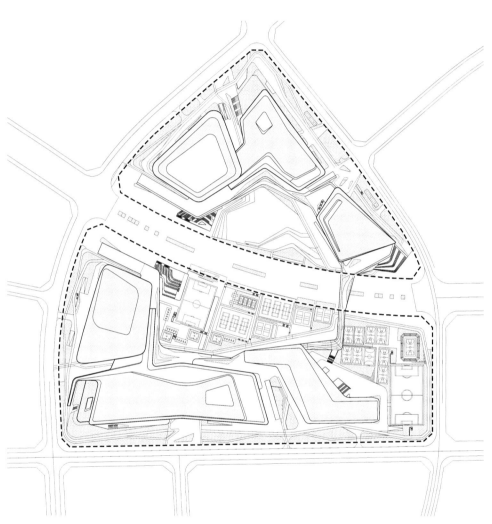

总平面图

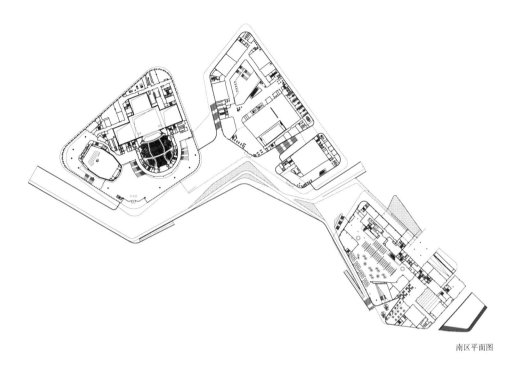

南区平面图

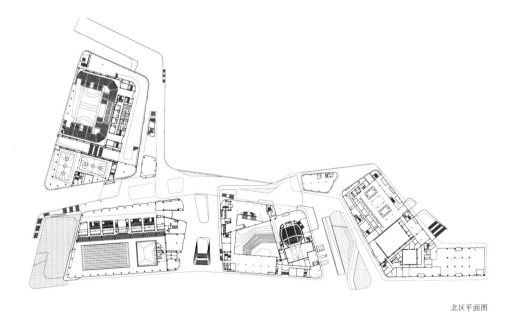

北区平面图

南区体育公园透视

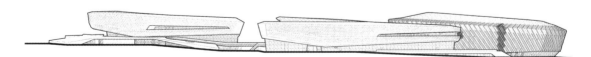

北区立面图

游泳馆室内透视

大剧院室内透视

大剧院前厅透视

大剧院室内透视

成都市双流区空港商务区展示中心
EXHIBITION CENTER OF CHENGDU SHUANGLIU

成都市双流区空港商务区展示中心（空港云）用地位于成都国际空港商务区，占地面积38180平方米。用地北侧紧临成双大道，面朝整个空港商务区且正对规划景观中轴，南侧紧邻双流运动公园，朝向双流国际机场，拥有开阔景观视野。展示中心由商务区建设发展场景体验馆和商业配套服务两大功能组成，总建筑面积12159平方米。

The Exhibition Center of Chengdu Shuangliu Airport Business District project (also known as Airport Cloud), covers an area of 38,180 square meters, closed to Chengshuang Avenue in the north, facing the entire airport business district and the central axis of the planned landscape. On the south side, it is adjacent to Shuangliu Sports Park and Shuangliu International Airport, with a wide landscape view. The exhibition center consists of two functions: the construction and development scene experience hall of the business district and the commercial supporting services, with a total floor area of 12,159 square meters.

空港"云"，整体标志性形态取意"云"，力图为紧临双流机场的空港商务区打造其专属的形象名片。空港商务区发展建设尚属起步阶段，用地四周荒地居多，大气整体的建筑体量能更好融入新区城市空间，带动周围建筑发展。同时，"云"的突出形态，也为从机场看过来的视野提供了视觉焦点，两者之间形成生动的视觉对话效应。

室内展厅虽是本项目设计的主体，但双流机场和空港新区发展盛况亦是展示中心希望呈现给参观者的景象。为此，内外场景的空间对话是展厅动线设计的主旨，整个游线、展示设计均考虑与城市对景充分结合，同时围绕"云中漫步"的核心主题，组织参观游览路线。

参观者穿过波光粼粼的水面，抵达展示中心主入口——圆形门厅，圆形门厅与环形水庭之间设有螺旋楼梯，引导参观者来到被抬高至11米标高的主体云朵展厅内，整个展陈设计结合外围城市风光与中央庭院景观，使参观者在观展行进过程中，仿若置身云端，既可观飞机起落盛景，又可在中央庭院小憩。

主展厅上方17米标高的夹层设有艺术沙龙及咖啡厅等功能，通过与平台屋面相接，参观者可驻足远眺商务区开发建设场景。参观结束后，来访者可通过弧形楼梯下至6米标高的裙房草坡屋面，继续游览草坡公园。

The Airport Cloud, whose overall iconic form "cloud", strives to create an exclusive image name card for the airport business district close to Shuangliu Airport. The Airport Business District is still in its initial stage of development and construction, surrounded by wasteland. The magnificent overall building volume can better integrate into the urban space of the new district and stimulate the development of the surrounding buildings. Meanwhile, the prominent form of the "cloud" also becomes a visual focus for the visual field seen from the airport, forming a vivid visual dialogue effect between the two.

Although the indoor exhibition hall is the main body of the project design, the grand development of Shuangliu Airport and Airport New Area is also the scene that the exhibition center hopes to present to the visitors. Therefore, the spatial dialogue between internal and external scenes is the theme of the contour line design of the exhibition hall. The whole visiting circulation and the display design are fully combined with the opposite view of the city. At the same time, the tour route is organized around the core theme of "walking in the cloud".

Visitors walk across the sparkling water and arrive at the main entrance of the exhibition center, that is the circular foyer. There are spiral staircases between the circular foyer and the ring water court to guide visitors to the main cloud exhibition hall raised to a height of 11 meters. The whole exhibition design combines the surrounding urban scenery and the central courtyard landscape, so that visitors can be in the process of watching the exhibition as if they were on the clouds. They can not only watch the ups and downs of the plane, but also have a panoramic view, while resting in the central courtyard.

The 17m-high interlayer above the main exhibition hall is equipped with art salon and cafe. By connecting with the platform roof, it enables visitors to stop and overlook the development and construction scene of the business district. After the visit, visitors can continue to explore the grass slope park through the curved stairs down to the grass slope roof of the 6m-high podium room.

从飞机鸟瞰建筑

规划设计在城市主干道一侧，通过景观草坡围合形成城市活力广场，北接空港景观轴，吸纳人气活力。为了规避北侧现状加油站的不利影响，用地北侧在靠近加油站的一角，通过景观造坡和植被栽种，尽可能降低加油站的视觉干扰，同时也将主要参观人流的进入动线往西侧景观河岸方向偏移，提升入口形象的同时也为营造更好的参观体验。

建筑局部覆土，与场地景观融合成统一的肌理，成为大地的景观，一汪浅浅的水景带来袅袅晨雾，风起时，其上的"云"也好似在涌动。

外立面设计结合二层景观大平台，通过横向金属百叶板强化建筑的水平延展特质，基于参数化设计对云朵图片进行的图案–数字转译，百叶板的出挑宽度相应形成丰富的参数渐变，在虚实交错之间，意图以抽象的建筑语言演绎"云朵"轻柔漂浮之势。横向金属板百叶同时减少太阳辐射热，提高了建筑的节能效率。

The planning is designed along the side city main Avenue, forming an urban vitality square surrounded by landscape grass slopes, and connecting the airport landscape axis in the north to absorb pedestrain vitality. In order to avoid the negative effects of the current gas station in the north side, the north side of the land is planted through landscape slopes and vegetation in the corner near the gas station to minimise the visual interference of the gas station. At the same time, the entry circulation of the main visitors is offset to the direction of the landscape river bank on the west side, which improves the image of the entrance and creates a better visiting experience.

The partial earth-covered construction and the site landscape integrate into a unified texture, becoming the landscape of the earth. A shallow waterscape brings a curling morning mist. When the wind blows, the "cloud" on it also seems to be surging.

The facade design combines the two-story landscape platform to strengthen the horizontal extension characteristics of the building through transverse metal shutters. Based on parametric design of cloud image pattern-digital translation, the unique width of the shutter corresponding form rich parameter gradient. Amid the alternation of virtuality and authenticity, it is intended to use the abstract architectural language to interpret the trend of "cloud" gently floating. The transverse metal plate louvers also reduce the solar radiation heat, and improve the energy saving efficiency of the building.

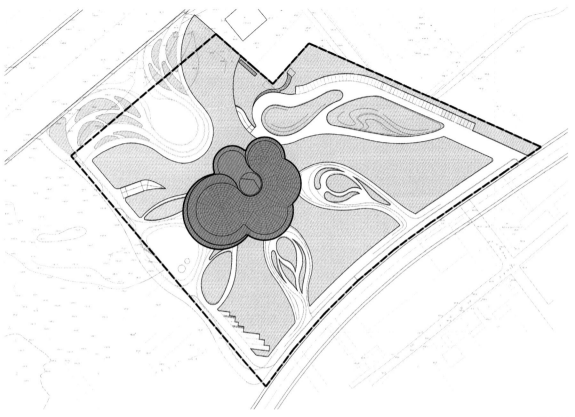

总平面图

　　为了强化二层主体展厅的漂浮感，底部架空结构采用化整为零的方式，所有结构钢柱分散布置，整个结构支撑体系的体量感被有意削弱，零零星星的结构杆件成为立面构成的一部分，彷如云朵下细细的雨丝。

　　建成开放后的"空港云"变成了当地人一处喜爱的休闲之地，承载着远眺观景、艺术体验、休闲互动、商务洽谈、亲子健身等活动，与互动观展本身，共同构成了空间体验营造的内容。我们希望，这个建筑是川味文化的载体，是寻常生活的载体，更是千万老百姓精神和记忆的载体。

In order to strengthen the floating feeling of main exhibition hall on the second floor, the overhead structure at the bottom is integrated into zero. All structural steel columns are scattered and arranged, thus the sense of volume of the whole structure support system is deliberately weakened. The sporadic structural members have become part of the facade composition, like thin rain filament under the clouds.

After the completion and opening, the "Airport Cloud" has become a favorite leisure place for local people, carrying the overlooking viewing, artistic experience, leisure interaction, business negotiation, parent-child fitness and other activities. Together with the interactive exhibition itself, it constitutes the content created by the space experience. We hope that this building is the carrier of Sichuan flavor culture, ordinary life, and the spirit and memory of tens of millions of people.

121

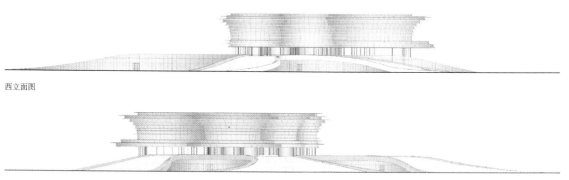

西立面图

东立面图

云朵下的庭院空间

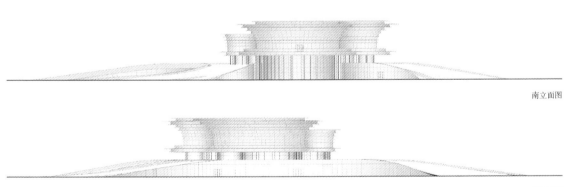

南立面图

北立面图

局部透视

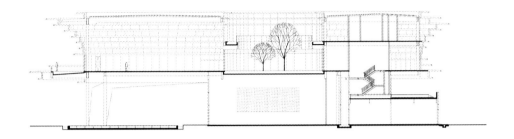

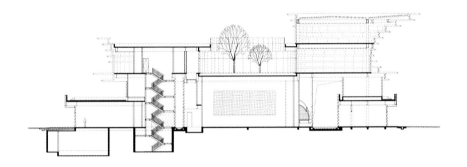

剖面图

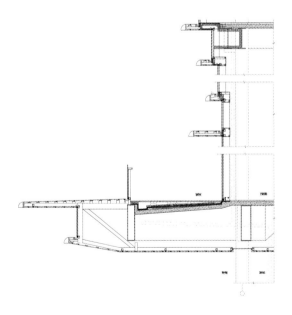

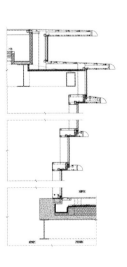

墙身节点

大邑长征文化园雪山剧场
THE SNOW MOUNTAIN THEATER OF DAYI LONG MARCH CULTURE PARK

长征文化园——"雪山剧场"位于成都市大邑县安仁镇，主体建筑是一幢拥有1200个座位的演艺剧场，建筑规模24000平方米，建成后这里将会上映一部以红军长征为背景的主题剧目——《长征》。设计之初，我们便在寻找一种表达方式。因为我们希望通过建筑与环境的打造，更好地还原当年红军长征的历史画面，让在此观看《长征》的观众能够以一种更为真切的感受快速融入剧目之中。

Located in Anren Town, Dayi County, Chengdu, the Snow Mountain Theater project has a 1,200-seat performing arts theater as its main building covering a floor of 24,000 square meters. After completion, there will be a theme drama released here with the background of the Long March of Chinese Red Army, that is, "The Long March". In the initial design, we were looking for a way to express it because we hoped to make use of the construction of the building and environment to better restore the historical picture of the Long March of the Red Army, so that the audience watching the "The Long March" here could quickly integrate into the play with a more real experience.

于是我们回顾历史，中国工农红军在四川境内共计转战了一年零八个月，留下了无数可歌可泣的悲壮故事。其中令人印象最为深刻的，便是著名的"爬雪山、过草地"。这早已成为红军长征的缩影，也成为红军长征最大无畏的精神写照。而项目本身位于成都市大邑县，著名的西岭雪山距此不远。于是，我们决定以具象的形式呼应"雪山"的环境，同时通过环境的打造映射"草地"，再造出红军长征经典的场景。

在建筑设计中，首先对雪山的形态进行了解构，提炼出"叠、切、折"三种建筑语汇。高低体量的叠加构建形态的整体秩序，切开完整的体形确立山形之间的起伏关系，顶部的弯折模拟群山的连绵，而立面的弯折则隐喻雪山独特的沟壑肌理。接着，对体量的高度进行了拆分控制，最终形成"主峰、次峰、群山"的三级构型。在建筑表皮的设计中，则以三角形单元对建筑形态进行拟合，采用白色的渐变彩釉玻璃，形成白色渐变的立面效果。夜晚时分，配合场地的泛光照明，白色的彩釉受光形成更加晶莹的效果。

Therefore, we looked back into the history and found that the Chinese Red Army of Workers and Peasants fought in Sichuan in a total of one year and eight months, leaving numerous tragic and stirring stories. Wherein, the most impressive story is the renowned "Climbing up the snow mountains and crossing the vast grasslands", which has been an epitome of the Long March as well as the spiritual reflection of the fearlessness of the Red Army during the Long March. And the project itself is located in Dayi County, Sichuan province, a place not far from the famous Xiling Snow Mountain, thus it gives us great design inspiraion.Finally, we decided to utilize the concrete form to echo the environment of the "snow mountain" and construct the environment to map the "grassland", thus recreate the typical scene in the Long March of the Red Army.

In terms of architecture design, firstly, we deconstructed the form of snow mountain and extracted three building languages, namely, "Die(piling), Qie(cutting) and Zhe (folding)". The superposition of high and low volumes built the overall order of the form, and the cutting of the building shape established the undulating relationship between the mountain forms. The bending at the top simulated the rolling of the mountains, while the bending on the facade implied the unique gully texture of the snow mountain. Then, we split and controlled the height of the volume, finally forming a three-level configuration of "main peak, secondary peak and mountains". In the design of the building facade, we fit the architectural form with the triangular unit, and adopted the white gradient colored glazed glass to form the facade effect of the white gradient. At night, with the flood lighting of the site, the white glaze will be given a more crystal light effect.

总平面图

由于该建筑是专门为《长征》"量身打造"的演艺剧场，因此还需要继续通过场地、建筑、室内的一体化设计，带给观众沉浸式的观演体验。于是，我们构建出一条"长征路"，贯穿建筑内外，采用"开端、发展、高潮、尾声"的线性叙事结构串联起众多观演的空间设计节点。

As the building is a specially "tailored" performance theater for the "Long March", we still need to generate an immersive performance experience to the audiences by virtue of the integrated design of the site, building and interior. In view of this, we built a "Long March Road", which runs through the inside and outside of the building, and adopted a linear narrative structure of "beginning, development, climax and end" to connect many viewing space design nodes.

　　从文化园入口的长征广场出发，三条通向长征剧场的栈道隐喻三路红军的长征路径，跨过前区草地后经由斜向坡道引至建筑主入口，随后进入光线逐渐变暗的入口门厅，肃静的环境让观众止步于此，缅怀烈士。随着观影流线继续前行，空间尺度会突然放大，进入了"雪山"休息厅，观众会在这里随着逐级上升的台阶体验翻越雪山的场景，最后到达明亮的休息厅，寓意长征结束后的红军迎来胜利曙光。

Starting from the Long March Square in the entrance of the Cultural Park, there are three plank roads stretching to the Long March Theatre, which metaphorically stand for the Long March paths of three branches of the Red Army. After crossing the front grassland, the slant slope will direct to the entrance of the main building. Then, there is the entrance vestibule as the lights darkening, and the solemn environment will stop audiences here to remember the martyrs. Moving forward following the viewing circulation, the space scale will be enlarged suddenly and lead the audiences into the "Rest Hall" of the "Snow Mountain" where they will experience the scene of climbing over the snow mountain with the ascending stairs before they finally reach the bright rest hall, which implies the bright victory dawn welcoming the Red Army after the Long March.

水面透视

　　待到观演结束，观众在归途中还会路过桤木河畔的追忆广场，在木栈道上回望剧场，仿佛也回溯了长征光辉的历史。

　　At the end of the viewing, the audiences will pass the Recalling Square by the Alder river and look back to the Theatre on the wooden plank road, as if also traced back to the glorious history of the Long March.

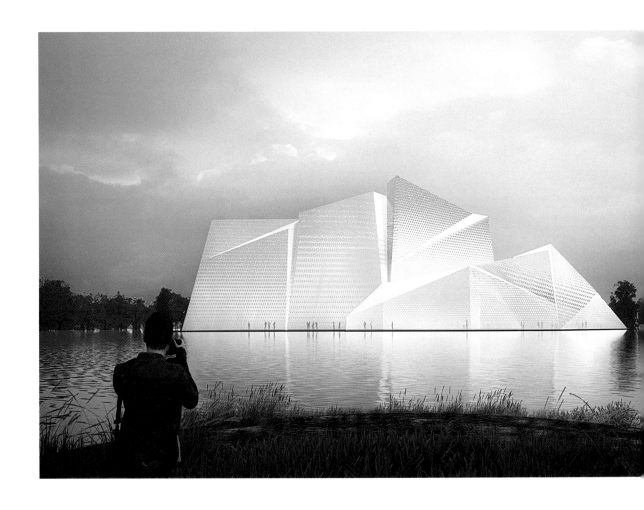

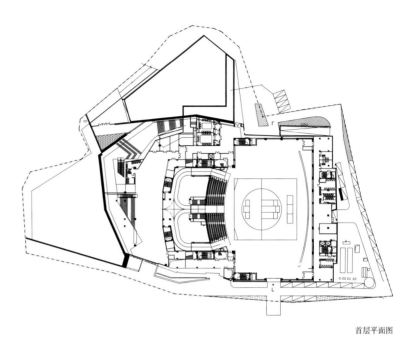

首层平面图

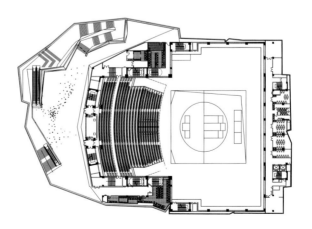

三层平面图

休息厅透视

室内台阶透视

纪念厅透视

中海成都天府新区超高层
ZHONGHAI CHENGDU TIANFU SUPERTALL

窗含西岭千秋雪，门泊东吴万里船。

在节奏飞快的当下时光里，几乎人人都把雪山当作深藏于心的神往之地。而在成都，推窗见雪山也许就在下一个清晨。中海成都天府新区超高层项目建成后，将成为世界上唯一一座能远眺海拔7000米以上雪山的480米以上超高层建筑。

The everlasting snow on West Mountain contains in my window, the outgoing ships for East Country gather around my gate.

In the f leeting current time when almost everyone regards the snow mountain as a place deep in the inward eye, in the future Chengdu enables you to bump into the snow mountains by simply opening the window just in the next morning. After completion, Zhonghai Chengdu Tianfu Supertall will become the only super high-rise building above 480 meters in the world which can see the snow mountain that above 7,000 meters.

塔冠透视

　　项目用地位于四川天府新区CBD核心区，距离天府广场27千米。项目由1号地的489米高的纯写字楼主塔和2号地的200米高的酒店配楼组成，总建筑规模约620000平方米。

　　The project plot is located in the CBD of Tianfu New Area, Chengdu, Sichuan Province, and is 27km away from Tianfu Square. The project includes a 489-meter tall office building in the plot 1, as well as a 200-meter tall hotel annex in the plot 2, with a total floor area of about 620,000 square meters.

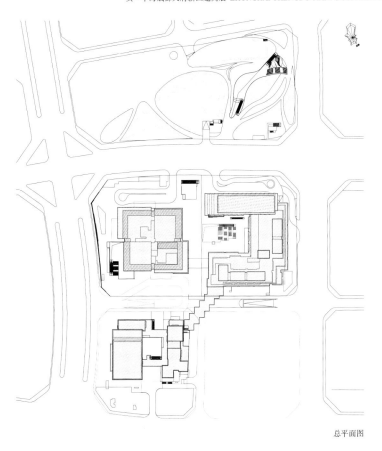

总平面图

　　设计团队从业主拿地开始就共同参与了项目的概念设计。方案最终由KPF中标，西南院和KPF、华艺共同组成的联合体则完成了整个项目的落地。实施方案中联合体团队延续并优化了中标方案"山巅、山峰、山麓、山脚的村落"的构思理念，将1号地的超高层塔楼比拟为"山峰"，超高层塔顶观光空间比拟为"山巅"；将2号地酒店比拟为"山麓"；1、2号地块间连续的商业空间则被比拟为"山脚的村落"，"山"的意向作为概念原型体现出了天府之国的文化和自然风光，将整座建筑"塑造"成为一座耸入云霄的山峰，完美融入了周边环境。

Our design team has been involved in the very beginning of the concept design since our client just acquired the plots. As a result, KPF won the competition, and the union of CSWADI, KPF and Hong Kong Huayi Design continued to complete the design and will accomplish its construction in the near future. In the implementation scheme, the union team continued and optimized the idea of the winning scheme, that is, the "mountain top, mountain peak, mountain foot, and mountain foot village", make the plot 1 main tower body as the "mountain peak", the sightseeing floor as the "mountain top", the hotel and apartment tower on the plot 2 as "mountain foot"; while the commercial space that connecting plot 1 and 2 as "mountain foot village". The idea of "mountain" reflects the cultural and natural scenery of Tianfu New Area and Chengdu, making the two towers literally a mountain that rises above the clouds, integrating the whole mountain into the surrounding environment.

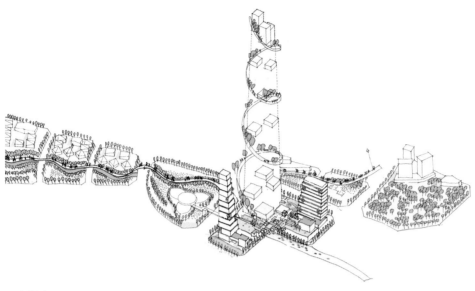

KPF 手绘设计理念

　　"山峰"超塔设计，根据空中大堂和配套设置需求在不同高度设置共享院落、景观退台、空中花园、屋顶庭院、天际花园等，形成沿主塔上升的自然景观，使办公人群体验到不同尺度的公园城市。

　　"山麓"酒店设计，配合酒店功能变化设置共享院落、景观退台等，使室外花园成为酒店空间的延展区，拓展酒店功能空间，增强酒店氛围。

　　"山脚的村落"商业设计，将建筑体量化整为零，结合不同的商业空间设置院落、退台、外摆区等，丰富商业空间氛围。

　　"山巅"观光设计，登高望远自古以来都映射着中国能人志士对人生理想抱负的追求，"会当凌绝顶，一览众山小"。在天府新区这个新兴都市的天际线，需要的也正是这样的一座承载了天府千年情怀的山巅。

　　In the design of the "mountain peak" super tall tower, the shared courtyard, landscape area, sky garden, roof courtyard, and sky gardens are set up at different heights according to the needs of the sky lobby and supporting facilities, forming a natural landscape rising along the main tower, so that the office people can experience the park city at different scales in a single building.

　　The "Mountain foot" hotel is designed with a shared courtyard, landscape steps, etc. combined with the hotel function changes, making the outdoor garden the extension area of the hotel space, expanding the hotel functional space, and enhancing the hotel atmosphere.

　　The commercial design of the "mountain foot village" integrates the building volume into parts, and sets up courtyards, setback terraces and external layout areas combined with different commercial spaces to enrich the commercial space atmosphere.

　　Then it is the sightseeing floor design of "Mountain peak". Since ancient times, the "mountain top" sightseeing reflected the pursuit of Chinese noble ideals and aspirations of life, "When shall I reach the mountain peak to belittle all the mountains in a glance". The skyline of Tianfu New Area, a new city area, needs exactly such a mountain top bearing the millennium feelings of Tianfu.

天府公园透视

　　超高层项目标志性强、建设成本高、参与方众多、建设周期长，是持续性的全过程设计。在实现过程中，也需要应对高压力的垂直交通设计、高效率的平面空间设计、高难度的消防设计、高精度的层高净高控制、高复杂性的TOD设计等关键技术难点的科学解答。我们的设计团队在会同联合体团队解读设计理念，打造本项目在形象表达上体现四川文化特色的同时，还针对项目的特殊技术难点，做了很多系统性的研究。我们希望这个项目不光是外在具备四川本土的文化传达，其内部的技术方案依然能体现出特别的川味"解答"。

The supertall project is highly iconic, with high construction cost, many participants and a long construction time cycle, which is a continuous whole-process design. In the process of implementation, it is also necessary to deal with the key technical difficulties such as high pressure vertical traffic design, high efficiency plane space design, difficult fire control design, high precision net height control, high complexity TOD design and so on. Our design team has done a lot of systematic research on the special technical difficulties of the project with other union design team, while also interpreting the design concept and creating the project to reflect the cultural characteristics of Sichuan in the image expression. We hope that this project not only has the external cultural expression of Sichuan, but also the internal technical solution can still reflect the special Sichuan style "solution".

天府公园透视

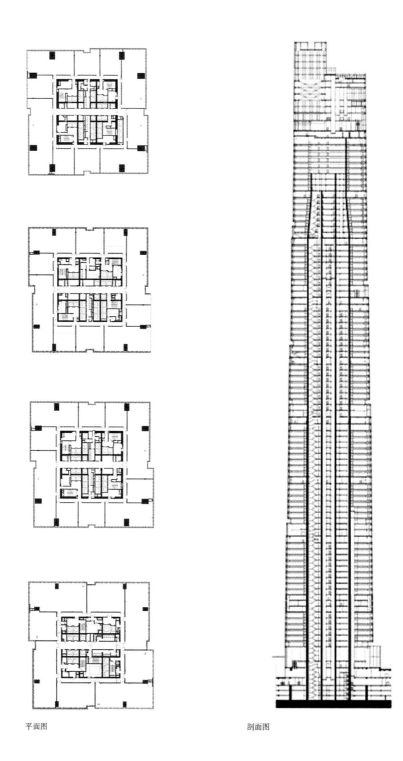

平面图

剖面图

本项目2号地的200米酒店配楼由我们的设计团队在概念方案的基础上完成全过程设计，并已经于2022年竣工。我们在紧凑的空间设计中，依然坚持在建筑贯穿200米的空间中出体现出地域特色。在建筑首层，设计通过富有四川味道的院落空间来组织酒店入口及商业空间。四川的砖瓦步道、"村口"的大树、青灰色毛石墙面、温润的木质格栅、连续的廊下空间，以及廊下潺潺的水瀑声，均含蓄地对来访者述说着传统的四川院落"安逸"的味道。休闲商业则通过连廊空间与院落连接，并结合退台设置外摆区，打造体验、情景式消费场所。以此复原出具备"川味"特色的市井生活场景。

The 200-meter hotel annex at plot 2 of the project has been completed in 2022 based on the concept scheme designed by our team. In the compact space design, we still insist on reflecting the regional characteristics of the building throughout the vertical space within 200 meters. On the ground floor, the hotel entrance and commercial space are organized through a Sichuan style courtyard space. From the brick and tile path of Sichuan, the big trees at the entrance of the village, the green and gray rubble stone wall, the warm wooden louvres, the continuous corridor space, and the murmuring waterfall below the corridor, all are telling the visitors the "comfortable" taste of the traditional Sichuan courtyard. The leisure commercial space is connected with the courtyard through the corridor space, and the outdoor terraces area is set up in combination with the setback platform to create an experiencing and spatial consumption place. In this way, the market life scene with the characteristics of "Sichuan style" is reappeared.

天府公园透视

空中平台透视

　　"道阻且长，行则将至，行而不辍，未来可期"，期待项目早日完工，重现"窗含西岭千秋雪，门泊东吴万里船"的怡然美景。

"Though the way is long and tough, as long as you travel, you can achieve; travelling without giving up, the future bright is coming". We are looking forward to the early completion of the project, so as to reproduce the leisure scene in the line "the everlasting snow on West Mountain contains in my window, the outgoing ships for East Country gather around my gate."

筑味 造味

空中平台透视

庭院透视

2020.12.3

拾年

一楼既立，百年不倒，时间让建筑变成了一所学校，领略艺术、回归本真，促进一代又一代人审美的提高；一时一日，与人相谐，时间造就了人间无数的相逢，公诸同好、知己在邻，让情感的寄托有所指向。

　　如此，根植土壤之艺术，人们相知之潜流，亦有大美可追。

2016.12.19

拾年・筑

2012

中物院成都科研创新基地科研综合楼
COMPLEX BUILDING OF CAEP IN CHENGDU

项目地点：四川·成都
项目规模：69900平方米
建成时间：2015

四川大学喜马拉雅文化及宗教研究中心
HIMALAYAN CULTURE AND RELIGION RESEARCH CENTER OF SICHUAN
UNIVERSITY

项目地点：四川·成都
项目规模：4300平方米
建成时间：2016

四川大剧院
SICHUAN GRAND THEATER

项目地点：四川·成都
项目规模：59000平方米
建成时间：2019

青白江文化体育中心
THE CULTURE AND SPORTS CENTER OF QINGBAIJIANG DISTRICT

项目地点：四川·成都
项目规模：98000平方米
建成时间：2016

2013

遂宁莲花会展中心
SUINING LOTUS EXHIBITION CENTER

项目地点：四川·遂宁
项目规模：154000平方米

泰达格调·青城岚田
TAIDAGEDIAO QINGCHENGLANTIAN RESIDENCE

项目地点：四川·都江堰
项目规模：78000平方米

成都星发工业配套生活服务区
CHENGDU XINGFA INDUSTRY LIVING QUARTERS

项目地点：四川·成都
项目规模：89000平方米
建成时间：2017

邛崃·上林郡
QIONGLAI SHANGLINJUN RESIDENCE

项目地点：四川·邛崃
项目规模：380000平方米
建成时间：2017

2014

遂宁老年大学及香林阁观音文化展示基地
SUINING UNIVERSITY FOR THE AGED AND XIANGLINGE GUANYIN CULTURE
DEMONSTRATION BASE

项目地点：四川・遂宁
项目规模：37000平方米
建成时间：2018

九寨沟丽思卡尔顿隐世精品度假酒店
A RITZ-CARLTON RESERVE HOTEL JIUZHAIGOU

项目地点：四川・阿坝
项目规模：35917平方米
合作单位：Wimberly Allison Tong & Goo 建筑事务所（新加坡）
建成时间：2022

重庆财富中心
CHONQING FORTUNE FINANCIAL CENTER (FFC)

项目地点：重庆
项目规模：135800平方米
建成时间：2017

高新区养老助残中心
HIGH-TECH ZONE OLD-AGED AND HANDICAPED SUPPORT CENTER

项目地点：四川・成都
项目规模：54000平方米

攀枝花・菩提苑
THE BODHI GARDEN, PANZHIHUA

项目地点：四川・攀枝花
项目规模：99188平方米

成都尧棠公馆
YIU TEUNG MANSION, CHENGDU

项目地点：四川·成都
项目规模：23748平方米
合作单位：株式会社日建设计
建成时间：2019

2015

西昌邛海泸山景区游客中心
TOURIST CENTER OF XICHANG QIONGLU SCENIC SPOT

项目地点：四川・西昌
项目规模：3000平方米
建成时间：2018

成都高新体育中心
CHENGDU HIGH-TECH ZONE SPORTS CENTER

项目地点：四川・成都
项目规模：201600平方米
建成时间：2021

成都图书馆、美术馆新馆
CHENGDU LIBRARY AND NEW ART MUSEUM

项目地点：四川・成都
项目规模：88100平方米

西藏林芝宾馆
LINZHI HOTEL, TIBET

项目地点：西藏・林芝
项目规模：25222平方米

海南陵水吉森度假酒店
HAINAN LINGSHUI JS HOLIDAY HOTEL

项目地点：海南・陵水
项目规模：95050平方米

2016

成都天府生物产业孵化园一期
CHENGDU TIANFU BIOLOGICAL INDUSTRY INCUBATION PARK PHASE I

项目地点：四川・成都
项目规模：660000平方米
建成时间：2019

西昌观海湾度假酒店
XICHANG LAKE-VIEW BAY RESORT HOTEL

项目地点：四川・西昌
项目规模：51000平方米

四川省妇女儿童中心
WOMEN AND CHILD CENTRE OF SICHUAN

项目地点：四川・成都
项目规模：54000平方米

援冈比亚国际会议中心
GAMBIA INTERNATIONAL CONFERENCE CENTRE

项目地点：冈比亚・班珠尔
项目规模：13800平方米
建成时间：2019

埃及国家会议中心
NATIONAL CONVENTION CENTER, EGYPT

项目地点：埃及・开罗
项目规模：167000平方米

埃及国家会展中心
NATIONAL CONVENTION AND EXHIBITION CENTER, EGYPT

项目地点：埃及·开罗
项目规模：354400平方米

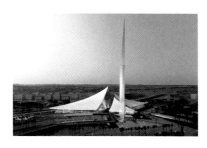

埃及国家歌剧院
NATIONAL OPERA HOUSE, EGYPT

项目地点：埃及·开罗
项目规模：108000平方米

2017

成都盛美利亚酒店
GRAN MELIA CHENGDU

项目地点：四川·成都
项目规模：66000平方米
建成时间：2020

成都"东来印象"文化体育中心
CULTURE AND SPORTS CENTER OF ORIENTAL IMPRESSION, CHENGDU

项目地点：四川·成都
项目规模：419900平方米
建成时间：2022

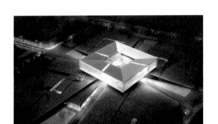

二里头夏朝遗址博物馆
ERLITOU SITE MUSEUM OF THE XIA CAPITAL

项目地点：河南·洛阳
项目规模：38600平方米

七彩云南·古滇名城
COLOURFUL YUNNAN ANCIENT DIAN TOWN

项目地点：云南·昆明
项目规模：1250000平方米
建成时间：2020

成都交投三岔湖酒店
CCIC SANCHAHU HOTEL

项目地点：四川·成都
项目规模：365579平方米

2018

宜宾华侨城超高层综合体
OCT YIBIN SUPER HIGH-RISE COMPLEX BUILDING

项目地点：四川·宜宾
项目规模：260000平方米

郑州滨河国际新城体育馆及图书馆
LIBRARY AND ARCHIVES OF INTERNATIONAL ECO-AQUAPOLIS, ZHENGZHOU

项目地点：河南·郑州
项目规模：135000平方米

中国核动力研究设计院设计研发大楼
CHINA NUCLEAR POWER RESEARCH AND DESIGN INSTITUTE

项目地点：四川·成都
项目规模：47389平方米
建成时间：2022

德格援藏干部交流中心
AID TIBET CADRES COMMUNICATION CENTRE OF DEGE

项目地点：四川·德格
项目规模：37900平方米

天府科创园超高层
TIANFU SCIENCE AND TECHNOLOGY SUPERTALL

项目地点：四川·成都
项目规模：537000平方米

2019

成都市双流区空港商务区展示中心
EXHIBITION CENTER OF CHENGDU SHUANGLIU

项目地点：四川 · 成都
项目规模：12000平方米
建成时间：2020

中海成都天府新区超高层
ZHONGHAI CHENGDU TIANFU SUPERTALL

项目地点：四川 · 成都
项目规模：517000平方米
合作单位：KPF、香港华艺设计顾问（深圳）有限公司

成都雅诗阁秦皇服务公寓
ASCOTT SERVICE APARTMENT CHENGDU

项目地点：四川 · 成都
项目规模：100300平方米
建成时间：2022
合作单位：KPF

成都空港国际会议中心
INDUSTRIAL SERVICE AREA OF PILOT FREE TRADE ZONE, CHENGDU

项目地点：四川 · 成都
项目规模：300000平方米

重庆礼嘉未来酒店
CHONGQING LIJIA FUTURE HOTEL

项目地点：重庆市
项目规模：47000平方米

重庆东站
CHONGQING EAST RAILWAY STATION

项目地点：重庆市
项目规模：340000平方米
合作单位：德国GMP事务所、中铁第四勘察设计院集团有限公司

四川攀枝花皇冠假日酒店
CROWNE PLAZA, PANZHIHUA, SICHUAN

项目地点：四川・攀枝花
项目规模：58600平方米

2020

天府农博园酒店
HOTEL OF TIANFU AGRI-EXPO GARDEN

项目地点：四川·成都
项目规模：46000平方米

深圳福田金融科技大厦
SHENZHENG INSTITUTE OF FINTECH RESEARCH CONSTRUCTION PROJECT

项目地点：广东·深圳
项目规模：77900平方米
合作单位：Zaha Hadid Architects 扎哈哈迪德事务所（伦敦）

四川名人馆
SICHUAN HALL OF FAME

项目地点：四川·成都
项目规模：51700平方米

顺德文化传播中心
CENTER FOR INTERNATIONAL CULTURAL COMMUNICATION, SHUNDE

项目地点：广东·顺德
项目规模：219600平方米

重庆站
CHONGQING RAILWAY STATION

项目地点：重庆
项目规模：1071500平方米
合作单位：中铁第四勘察设计院集团有限公司

祠堂街历史街区改造
REDEVELOPMENT OF CITANGJIE

项目地点：四川·成都
项目规模：33000平方米

2021

西南交通大学大学生创新创业教育中心
INNOVATION AND ENTREPRENEURSHIP EDUCATION CENTER FOR COLLEGE
STUDENTS OF SOUTHWEST JIAOTONG UNIVERSITY

项目地点：四川·成都
项目规模：29000平方米

大邑长征文化园雪山剧场
THE SNOW MOUNTAIN THEATER OF DAYI LONG MARCH CULTURE PARK

项目地点：四川·成都
项目规模：23770平方米

沈阳西华北里盛京皇城剧场概念设计
CONCEPT DESIGN OF XIHUABEILI THEATER, SHENYANG

项目地点：辽宁·沈阳
项目规模：15000平方米

成都天府国际生物制药产业加速器四期
TIANFU INTERNATIONAL BIOPHARMACEUTICAL INDUSTRY ACCELERATOR
PHASE IV, CHENGDU

项目地点：四川·成都
项目规模：258700平方米

郑州国际会议中心
ZHENGZHOU INTERNATIONAL CONVENTION CENTER

项目地点：河南·郑州
项目规模：130168平方米

郑州奥体国宾酒店
ZHENGZHOU OLYMPIC SPORTS HOTEL

项目地点：河南・郑州
项目规模：83441平方米

成都天府奥体公园国际会都岛酒店
CHENGDU TIANFU OLYMPIC PARK INTERNATIONAL CONVENTION ISLAND
HOTEL

项目地点：四川・成都
项目规模：67100平方米
合作单位：艾奕康设计与咨询（深圳）有限公司、清华大学建筑设计研
　　　　　究院有限公司

大理规划馆、美术馆及会议中心
DALI COUNTY PLANNING EXHIB HALL ART GALLERY AND CONVENTION
CENTER

项目地点：云南・大理
项目规模：100000平方米

苏州北站
SUZHOU NORTH RAILWAY STATION

项目地点：江苏・苏州
项目规模：1150000平方米
合作单位：中国建筑设计研究院有限公司、中铁第五勘察设计院集团有
　　　　　限公司

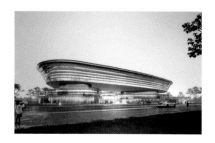

四川天府新区市民之家
CIVIC CENTER, TIANFU NEW AREA, SICHUAN

项目地点：四川・成都
项目规模：83000平方米

成都天府生物产业孵化园三期
CHENGDU TIANFU BIOLOGICAL INDUSTRY INCUBATION PARK PHASE III

项目地点：四川 · 成都
项目规模：113000平方米

天府国际健康服务中心
TIANFU INTERNATIONAL HEALTH SERVICES CENTER

项目地点：四川 · 成都
项目规模：430930平方米
建成时间：2022

2022

成都2024世园会瀑布酒店
WATERFALL HOTEL, 2024 CHENGDU WORLD HORTICULTURAL EXPOSITION

项目地点：四川・成都
项目规模：33000平方米

郫县豆瓣博物馆
PIXIAN CHILI BEAN SAUCE MUSEUM

项目地点：四川・成都
项目规模：15000平方米

德阳经开区群众文体中心
PEOPLE'S CULTURE AND SPORTS CENTRE,JINGKAI DISTRICT, DEYANG

项目地点：四川・德阳
项目规模：26000平方米

成都天府国际生物制药产业加速器三期
TIANFU INTERNATIONAL BIOPHARMACEUTICAL INDUSTRY ACCELERATOR PHASE III, CHENGDU

项目地点：四川・成都
项目规模：250000平方米

建川电影博物馆聚落・建昌博物馆
JIANCHANG MUSEUM, JIANCHUAN FLIM MUSEUM CLUSTER

项目地点：四川・西昌
项目规模：12000平方米

彭州小鱼洞温泉酒店
XIAOYUDONG ONSEN RESORT & SPA, PENGZHOU

项目地点：四川·彭州
项目规模：6500平方米

重庆大学工科实验楼
CHONGQING UNIVERSITY SCIENCE HUB

项目地点：重庆
项目规模：72000平方米

成都天府生物产业孵化园M地块
TIANFU INTERNATIONAL BIOLOGICAL CITY SITE BLOCK M

项目地点：四川·成都
项目规模：97270平方米

绵阳剧场
MIANYANG THEATER

项目地点：四川·绵阳
项目规模：85000平方米

·图片提供：存在建筑　404NF STUDIO
蒋人可　杨　荣　燕　飞

2017.01.14

拾年・人

2012

郑　勇
刘作卓
肖迪佳
吴玉婧
汪　宇
沈茂良
屈国伟
贾　伟
龚　良

2013

郑　勇
刘作卓
肖迪佳
吴玉婧
汪　宇
沈茂良
李涵博
贾　伟
陈嘉乐
陈艳妍
刘运娜
龚　良

2014

郑　勇
刘作卓
肖迪佳
吴玉婧
汪　宇
沈茂良
仲泉丞
贾　伟
李涵博
刘运娜
陈嘉乐
陈艳妍
龚　良

2015

郑　勇
刘作卓
肖迪佳
吴玉婧
汪　宇
沈茂良
仲泉丞
刘　汉
李涵博
刘运娜
陈嘉乐
陈艳妍
李知浩
龚　良

2016

郑　勇
刘作卓
肖迪佳
吴玉婧
汪　宇
金　鑫
仲泉丞
刘　汉
李涵博
刘运娜
张　涵
陈嘉乐
陈艳妍
尹　涛
张　琪
周凌寒

2017

郑　勇
刘作卓
肖迪佳
张皓森
刘　宇
金　鑫
仲泉丞
刘　汉
李涵博
刘运娜
林少康
陈嘉乐
陈艳妍
张　涵
尹　涛
张　琪
周凌寒

2018

郑　勇
刘作卓
肖迪佳
张皓森
刘　宇
金　鑫
仲泉丞
刘　汉
李涵博
刘运娜
金秋平
严星瑶
陈嘉乐
尹　涛
陈艳妍
李永涛
张　涵
吴明奇
张　琪
周凌寒

2019

郑　勇
刘作卓
肖迪佳
李国熊
刘　宇
金　鑫
仲泉丞
刘　汉
李涵博
刘运娜
尹　涛
吴明奇
陈嘉乐
赵炎鹏
陈艳妍
袁丹龙
郑　珣
金秋平
李永涛
张　涵
严星瑶

2020

郑　勇
刘作卓
肖迪佳
李国熊
刘　宇
仲泉丞
李涵博
刘　汉
陈嘉乐
陈艳妍
郑　珣
金秋平
贾一凡
续文琪
刘运娜
罗力铭
严星瑶
童　骞
赵炎鹏
李永涛
尹　涛
吴明奇
张　涵
彭　一

2021

郑　勇
刘作卓
肖迪佳
李国熊
刘　宇
李涵博
陈嘉乐
仲泉丞
刘运娜
张　涵
熊　雪
郑　珣
杨　伊
续文琪
陈艳妍
米佳锐
吴明奇
贾一凡
童　骞
赵炎鹏
罗力铭
尹　涛
金秋平
彭　一
徐铎轩
杨　屹
姚　汉
陈　冉

2022

郑　勇
刘作卓
仲泉丞
李国熊
刘　宇
陈嘉乐
刘运娜
李涵博
陈艳妍
尹　涛
杨　屹
杨　伊
佘杏奕
张入介
张　涵
陈志强
罗力铭
罗晓梦
姚　汉
续文琪
彭　一
童　骞
熊　雪

178